Peter Johnston Freyer

The Modern Treatment of Stone in the Bladder by

Litholapaxy

Peter Johnston Freyer

The Modern Treatment of Stone in the Bladder by Litholapaxy

ISBN/EAN: 9783337811754

Printed in Europe, USA, Canada, Australia, Japan

Cover: Foto ©berggeist007 / pixelio.de

More available books at **www.hansebooks.com**

THE

MODERN TREATMENT

OF

STONE IN THE BLADDER

BY

LITHOLAPAXY

THE

MODERN TREATMENT

OF

STONE IN THE BLADDER

BY

LITHOLAPAXY

A DESCRIPTION OF THE OPERATION AND INSTRUMENTS
WITH CASES ILLUSTRATIVE OF THE DIFFICULTIES
AND COMPLICATIONS MET WITH

BY

P. J. FREYER, M.A., M.D., M.CH.

BENGAL MEDICAL SERVICE; CIVIL SURGEON, MUSSOORIE

LONDON

J. & A. CHURCHILL

11 NEW BURLINGTON STREET

1886

Dedicated

TO

ROBERT McDONNELL, M.D., F.R.S.

PRESIDENT OF THE ACADEMY OF MEDICINE IN IRELAND

LATE PRESIDENT, COLLEGE OF SURGEONS, IRELAND

SURGEON TO STEEVENS' HOSPITAL, DUBLIN

ETC. ETC.

IN GRATEFUL RECOLLECTION OF MANY PRACTICAL LESSONS IN SURGERY

AND NUMEROUS ACTS OF KINDNESS

CONFERRED BY HIM

ON HIS FORMER PUPIL AND HOUSE SURGEON

THE AUTHOR

PREFACE.

—•—

THE rapidity with which a small pamphlet of mine on
" Litholapaxy," recently published in India, and which
was mainly a reprint from the *Lancet*, is being disposed
of, shows that the subject dealt with is now eliciting
considerable attention. I have, therefore, determined
on reproducing that pamphlet in the present form.

In this little work, I aim at placing before my professional brethren a guide to the modern treatment of
stone by litholapaxy, of a thoroughly practical nature.
In addition to a detailed description of the operation
and instruments employed, I have devoted particular
attention to the difficulties and complications met with,
and illustrated these. and the best means of dealing
with them, by cases. in detail, from my own practice.
now extending over 128 cases of the operation.

I have also elaborated the new method of diagnosis
of stone recently brought by me to the notice of the
profession.

The very favourable manner in which my work in litholapaxy has, from time to time, been noticed by the Medical Press and leading medical men at home, in America, and in India encourages me in the hope that this monograph may contain some original observations of practical interest to my fellow-labourers, especially those serving in India.

CONTENTS.

INTRODUCTORY.

From time to time, during the past four or five years, papers of mine have been published in the medical journals,* giving full details of several series of litholapaxy operations performed by me.

Having now completed 128 cases of the operation, I propose dealing with them comprehensively in the present monograph, repeating some of the remarks recorded in the papers above alluded to, and adding many further observations of a practical nature which I have learnt from an increased experience of the operation. I propose further entering into a detailed description of the operation and the various modern instruments employed in its performance, adding a series of cases from my own practice illustrative of the chief difficulties met with in the operation, and the means by which these difficulties may be overcome. It is hoped that the record of a large number of cases of this operation, from the practice of a single operator, may prove interesting to the profession at large, and that the results obtained may have some effect in bringing into more general practice, amongst my professional brethren in this

---

* *Indian Medical Gazette*, Dec. 1882, Feb. 1883, March 1884, April 1885, Jan. 1886, Feb. 1886; the *Lancet*, Feb. 28, March 7 and 14, 1885.

country, an operation which, though still in its infancy, has, undoubtedly, a brilliant future in store for it.

The extreme aversion with which the natives of India regard any method of treatment that involves several distinct surgical proceedings, extended over an indefinite period, is well known to every surgeon who has practised in this country. The knowledge of this fact alone, putting aside altogether the comparative merits of lithotrity and lithotomy, was sufficient to deter surgeons from practising the old operation of lithotrity. And, as the result, though there were between two and three thousand operations for stone performed annually in Indian hospitals, I may confidently say that the number of lithotrity operations undertaken might be counted on the fingers of one's hands. An operation which in England and Europe generally, in America and the Colonies, though periodically brought to the front by some brilliant operator, can scarcely be said to have held its own against lithotomy amongst the profession generally, in India was never destined to play an important part, owing to the peculiar temperament of the natives above alluded to.

When, however, some six years ago, Professor Bigelow, of Harvard, U.S., introduced his new operation of "Litholapaxy" to the notice of the profession, and, scattering to the winds all previously held theories as to the deleterious effects of prolonged instrumentation in the bladder, demonstrated by a series of successful cases that, instead of subjecting the patient suffering from stone to a number of short sittings, extending over an indefinite period, the calculus might be crushed and evacuated at one sitting, the chief objection to the crushing operation, so far as this country was concerned, seemed to have been removed.

As with many of my professional brethren in India, lithotomy in my hands had proved a fairly successful operation. I did not, therefore, at first abandon the cutting for the new operation. I must confess, however, that the main cause of my hesitation in adopting the new operation was the depreciative manner in which it was criticized by Sir Henry Thompson on its introduction. I need scarcely say that English surgeons have been in the habit of receiving as almost equivalent to divine law the utterances of that distinguished surgeon on any point connected with the surgery of the urinary organs. When, therefore, I read of the "disastrous" results that he anticipated from Bigelow's operation, I naturally hesitated in adopting it. And it was not till I had subsequently read of the brilliant results Sir Henry himself obtained from that operation with reference to which he had uttered such gloomy forebodings, that I finally decided on giving the operation a trial.

Shortly after the above lines were first published in the *Lancet* * I was favoured by a very friendly letter from Sir Henry Thompson, in the course of which—referring to my article in that journal—he says : " I felt much pained at your belief that I had opposed the operation at the outset, or foretold it would be ' disastrous.' All I can say is, that I have no recollection of having done so." And again : " I expressed my dislike to Bigelow's

* Feb. 28, 1885.

instruments, a dislike I still maintain ; but his method
I have never opposed, and, since trying, have never
failed to espouse and to praise."

Now, I should be extremely sorry to misrepresent Sir
Henry Thompson, to whom, through his writings and
lectures, I owe more than I do to any other surgeon
in that branch of surgery to which he has particularly
devoted himself. But, after a re-perusal of his
utterances on the subject, I must say that his writings,
which on other subjects are so lucid, on this render him
liable to misinterpretation. I endeavoured in the latter
part of my article to render the position assumed by
Sir Henry, with reference to Bigelow's operation, clear,
by quotations from his own writings ; but the editor
of the *Lancet*, after keeping my paper unpublished
for months, emasculated and altered it in such a
manner that, on publication, it was scarcely known to
the author. In the action of the *Lancet* in delaying
the publication of my article, I suppose I have no
special cause of complaint, for I believe it is the policy
of that journal to ignore everything " Indian." But
when my paper was published, I might, in fairness, have
expected that statements made by me should not have
been left unsubstantiated by the omission of important
passages from it. That my paper was a substantial
contribution to the literature of a subject comparatively
new to the profession I have the satisfaction of know-
ing, from the favourable notice of it that has been
taken by the medical press and several leading medical
men.

To return from this digression, I shall reproduce later

on the quotations from Sir Henry Thompson's writings above alluded to, and the reader will have an opportunity of judging for himself the meaning they convey. I cannot, with Sir Henry, disassociate Bigelow's operation and the instruments employed. It reminds one very forcibly of the play of " Hamlet " without the leading character. It could scarcely have been expected that the instruments should be perfect on their first introduction, nor are they perfect now, though vast strides in this direction have lately been made. particularly by Sir Henry Thompson and Bigelow himself.

With these introductory remarks, I shall now proceed to describe in detail the instruments required in the operation of litholapaxy.

THE INSTRUMENTS EMPLOYED IN LITHOLAPAXY.

To appreciate fully the great revolution in the surgery of the bladder involved in Bigelow's operation, it will be necessary to give a brief sketch of the history of lithotrity; and this will involve a description of the development of one of the essential instruments of the modern operation—viz., the lithotrite.

Though cutting operations of various kinds, for stone, have been practised from the earliest ages, it was not till the beginning of the present century that the idea of removing a stone by pulverizing it within the bladder, and allowing the débris to escape by the natural passages, was entertained. The first to crush a stone on scientific principles was the great French surgeon Civiale, in 1824. This he effected by an instrument named the "trilabe," a species of drill consisting of a central axis and three claws, which, after introduction of the instrument into the bladder through the urethra, were made to project and catch the stone. The reduction of the stone to fragments was effected by drilling holes in it in various directions, till it crumbled into débris. The operation, which he named "lithotrity,"

extended over several sittings, the fragments passing
away naturally with the urine. It will be observed
that the disintegration of the stone was accomplished
by a drilling rather than by a crushing process.
Shortly afterwards a great improvement was effected
by Weiss, of London. He constructed an instrument
by which the stone was grasped between two short
blades, bent at an angle with the shaft, and reduced
to fragments by a true crushing process. And, though
various improvements have since been effected, it may
be said that this was the model on which all modern
lithotrites are constructed.

In the development of the lithotrite, various changes
have been made in the method of applying the motive
power by which the stone is crushed between the
blades. Thus, in Heurteloup's time, the patient was
placed in a peculiar shaped bed to which a vice was
attached. After the lithotrite was introduced into the
bladder, and the stone grasped between its jaws, it
was fixed in the vice, and the stone reduced to débris
by blows of a hammer applied outside. Some time
after, a great improvement was effected, in the introduction of the screwing process by Hodgson, of Bir-
mingham. Sir William Fergusson advocated the rack-
and-pinion system in the lithotrite ; but the screwing
process is that generally now adopted. Lastly, Sir
Henry Thompson adapted the cylindrical handle to the
lithotrite ; and this, combined with Weiss's method,
by which the sliding action may be converted into a
screwing one, is now generally employed in the construction of modern lithotrites.

Turning aside now from the instruments, let us glance
at the principles involved in the operation. As already
mentioned, Civiale's practice was to crush small
quantities of stone at repeated and frequent sittings,
each extending over a few minutes only, the detritus
coming away by natural efforts with the urine. From
time to time, however, attempts were made to assist
Nature in getting rid of the débris by artificial means.
For this purpose currents of water, injected into the
bladder through a large catheter from a syringe, were
employed by Heurteloup and others. In 1846, Sir

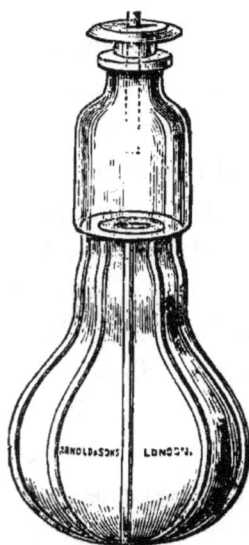

FIG. 1.

Philip Crampton, of Dublin, invented a suction apparatus resembling a large soda-water bottle with
a tap at the neck, which was exhausted of air and then applied to
a catheter previously introduced
into the bladder, and, in this way,
an attempt made to get rid of
the fragments. Subsequently, Mr.
Clover designed his syringe (Fig. 1),
which consisted of an india-rubber
bulb with a glass receiver, from
which water was pumped into,
and withdrawn from, the bladder
through a catheter, No. 12 or 13 ;
and, in this way, a certain quantity of sand brought
away. Then, again, Sir William Fergusson endeavoured
to complete the operation at one sitting by withdrawing the fragments through the urethra by means of
long and slender lithotrites. Sir Henry Thompson,

though apparently adverse to this method at first,* subsequently employed it for a time.† But it came to be regarded by the profession as a very dangerous process, often inflicting severe injury on the urethra.

All these methods of artificial evacuation of débris were invented with a view to obviate the recognized danger of allowing rough and sharp fragments of stone to remain in the bladder—a common cause of cystitis. Each method enjoyed a temporary, though transient, notoriety; but they one and all fell into disrepute, for the simple reason that they failed to accomplish the object at which they aimed, and, at the same time, caused a great deal of irritation. And the practice which Civiale had inculcated, of short and frequent sittings, the débris being allowed to come away by natural efforts, came eventually to be recognized, by universal consensus of opinion amongst the profession, as the most safe and judicious.

Such was the position of lithotrity in 1878, when Bigelow appeared on the scene with his new operation, and proposed to revolutionize the whole system by crushing and evacuating the stone at one sitting, no matter how prolonged, and no matter how large the stone might be, provided only that it was capable of being grasped and crushed by the large lithotrites then proposed.

Bigelow's operation practically resolves itself into two

* "Lectures delivered at the College of Surgeons, England, 1884," by Sir H. Thompson, p. 117.

† "Lithotrity at one or more Sittings" (*Lancet*, vol. i. 1879, p. 145).

proceedings—the reduction of the stone to fragments and the evacuation of the débris from the bladder.

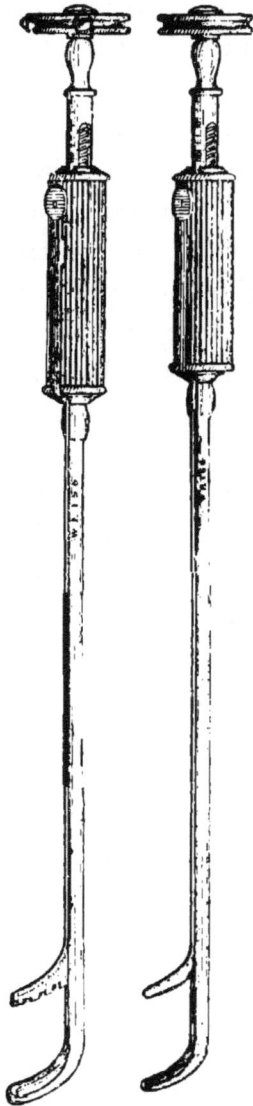

The crushing of the stone is accomplished by means of lithotrites similar to those employed for the old operation of lithotrity, except that, owing to the increased scope of the new operation in dealing with large and hard calculi, some of the lithotrites are constructed much larger and stronger than those formerly in use. In Figs. 2 and 3 are illustrated lithotrites constructed on the well-known model of Weiss and Thompson, and admirable instruments they are. They possess the cylindrical handle introduced by Sir Henry Thompson, which (in the words of the distinguished inventor) "enable you, in the search for a small stone or fragments, to execute rapid and delicate movements which would be impossible in an instrument without the cylindrical handle." They also possess the new mode of changing sliding into screwing action, and *vice versâ*, introduced by Messrs. Weiss & Son. When the small button in front of the cylinder is pushed back into the position indicated in the illustra-

Fig. 2.   Fig. 3.

tions, the instrument is " locked," and then the male blade moves within the female blade by a screwing action only ; but when the button is pushed forward in the direction of the blades, the instrument is " un-locked," and the screwing is converted into a sliding action.

For the operation of litholapaxy, three varieties of this instrument are generally employed :—(1) A large fenestrated lithotrite (Fig. 2) for crushing large and hard stones. In this the male blade, which is deeply serrated or toothed, passes through the female blade, driving the débris through the opening in the latter, or tossing it away on either side, so that no blocking of the blades by fragments can occur. (2) A medium-sized instrument of the same description as the last, for dealing with stones of moderate size, or for use when the urethra is not sufficiently capacious to admit the larger lithotrite. (3) A flat-bladed, non-fenes-trated lithotrite (Fig. 3), which is used for reducing fragments into fine powder, after the coarse work of break-ing up the stone has been ef-fected by means of the fenes-trated instruments. This may also be used with advantage in crushing very small, and especially soft, stones.   Fig. 4 illustrates the blades of

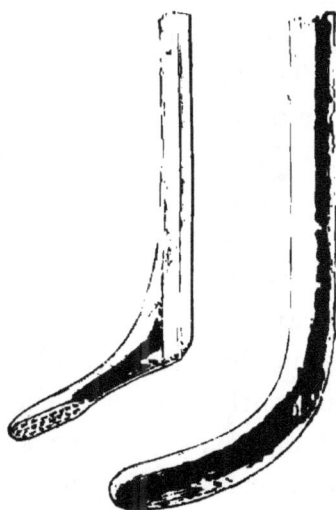

Fig. 4.

another variety of flat-bladed lithotrite that is sometimes employed, in which there is a small opening at
the heel of the female blade. For some time after
I began to practise litholapaxy, I used these flat-
bladed instruments a great deal ; but I now, as a rule,
complete the operation by means of the fenestrated
lithotrites alone. When large evacuating catheters
can be passed, as in the great majority of cases, it is
unnecessary to reduce the stone to fine sand, as coarse
débris can pass through these tubes into the aspirator,
and it is a waste of time to reduce the débris to a finer
consistence than what will pass through the canulæ
with facility. Cases will, however, occur in which,
from various causes, small canulæ only—say, Nos. 13
or 14—will pass through the urethra ; and in such cases
the flat-bladed lithotrite will be of great advantage in
pulverizing the fragments into fine sand.

The flat-bladed, non-fenestrated instruments possess
the disadvantage of being liable to get clogged with
débris. Sir Henry Thompson says : * " A collateral
advantage of this flat-bladed instrument is, that it will
hold a good deal of débris without undue augmentation
of its size, so that not a little can be safely brought
away by the urethra, if desired, whenever the instrument is withdrawn." In writing thus, Sir Henry
evidently confounds the old and new operations. " We
should," as Bigelow says, " distinctly recognize that
what can be withdrawn in a lithotrite could come better
through a tube, and that the only province of the

* " Diseases of Urinary Organs," sixth edition, 1882, p. 78.

lithotrite should be to pulverize, or, indeed, merely comminute, and not to evacuate."

In Bigelow's lithotrite (Fig. 5), the cylindrical handle of Thompson's instrument is retained for the left hand ; but, for the wheel for the right hand, a ball is substituted. This is an undoubted improvement, affording a much firmer purchase—a point of great importance when dealing with a large and hard calculus.   But the special feature of Bigelow's lithotrite is the introduction of a new mode of locking the instrument.   This is effected simply by a quarter rotation of the right wrist, whilst the hands are in position, without any displacement of the fingers; whilst a quarter rotation of the wrist in the opposite direction unlocks the instrument.   In the lithotrite of Weiss and Thompson, the thumb of either hand has to be disengaged to move the button, a performance which tends to render the lithotrite in the bladder unsteady at the critical moment of catching the stone. By the ingenious device of Bigelow this objection is obviated—a decided improvement. On the whole, the movements of this lithotrite are easier and more graceful than in any instrument I have ever worked with.   So much for the handle of Bigelow's lithotrite.

I cannot say, however, that I like the blades of this instrument nearly so well as those of the fenestrated lithotrites by Weiss

FIG. 5.

and Thompson already described. "The blades of this lithotrite consist of a shoe, or female blade, the sides of which are so low that a fragment falls upon it; while the male blade, or stamp, offers a series of alternate triangular notches by whose inclined. planes the detritus escapes laterally after being crushed against the floor and rim of the shoe. At the heel of the shoe, where most of the stone is usually comminuted, and where the impact is therefore greatest, the floor is high and discharges itself laterally, while its customary slot is made to work effectually" (Bigelow). The blades are essentially non-fenestrated, and liable to get clogged with débris, as I have frequently found in practice, and, therefore, objectionable when dealing with large calculi.

By a combination of the handle and locking action of Bigelow's lithotrite with the fenestrated blades of that by Weiss and Thompson, a more perfect instrument might, I think, be produced; and one of this description is now being constructed for me by Messrs. Weiss & Son, of London.

The second object aimed at in the operation—the removal of the débris from the bladder—is accomplished by means of large cylindrical tubes, or evacuating catheters, introduced through the urethra, and an aspirator, or suction apparatus, attached thereto.

Some time before the introduction of Bigelow's operation, it had been demonstrated by Otis, of New York, that the urethra in the adult male is much more capacious than had previously been imagined; and this discovery undoubtedly paved the way towards the development of the new operation. The canulæ employed vary in size from Nos. 14 to 20, English scale, according to the capacity of the urethra. In my own practice I have not found it necessary to use a larger canula than No. 18; and through a tube of this calibre I have removed the débris of calculi weighing $3\frac{1}{4}$ ounces. I have, however, met with cases in which a No. 19 or 20 canula might have been passed with facility. Canulæ are made of thin silver, and vary in shape, some being straight, and some slightly curved at the extremity (Figs. 6, 7. 10). The latter I prefer, as I find them more easy of introduction. The orifice, or eye, should be large enough to admit any fragment that will pass through the tube.

Though the evacuating catheters remain much the same now as on their introduction by the distinguished originator, several modifications have been effected in the aspirator; and these I shall now describe, indicating the varieties of aspirator that I have found most effective.

The original aspirator of Bigelow, with curved canula attached, is represented in Fig. 6. It consists of

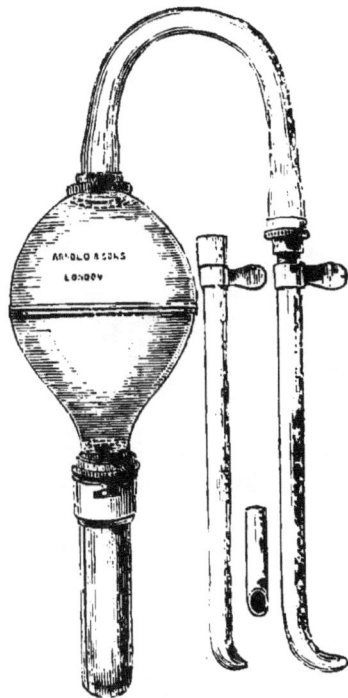

FIG. 6.

of an elastic bulb, or central portion, to the lower
extremity of which is attached a removable cylindrical
glass-receiver; whilst from its upper part passes an
india-rubber tube, the end of which fits on to the
evacuating catheter previously introduced into the
bladder.   The apparatus, previously filled with water,

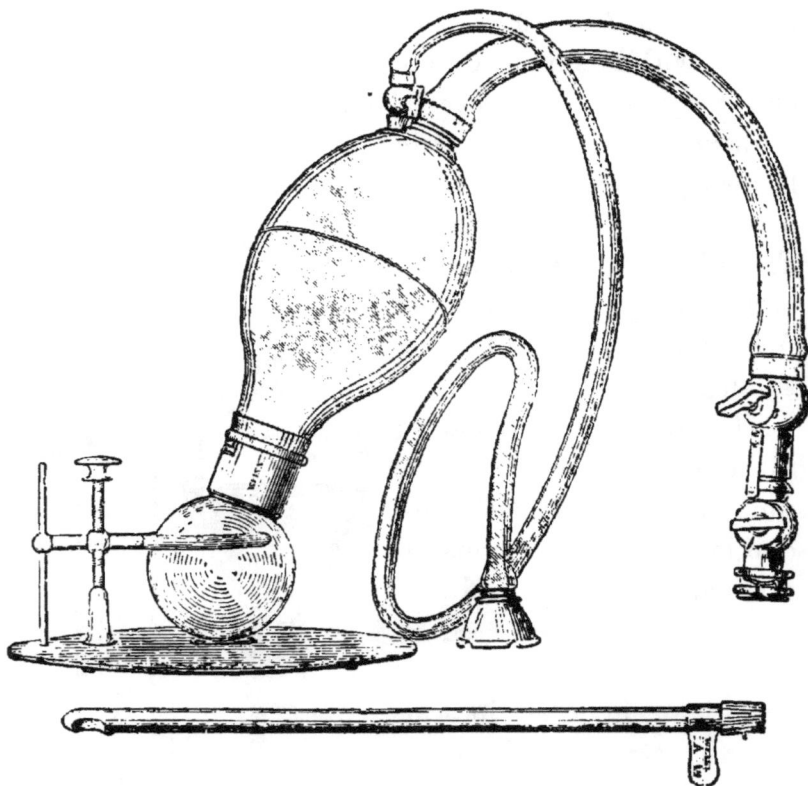

FIG. 7.

acts as a kind of syphon.   By alternate compression
and expansion of the bulb, the water is pumped into,
and withdrawn from, the bladder, and the débris,

which is carried back into the aspirator, falls down into the glass-receiver, and is there retained.

A more recent form of Bigelow's aspirator is represented in Fig. 7, resting on a stand; and to which is added an elastic tube, or hose, provided with a stop-

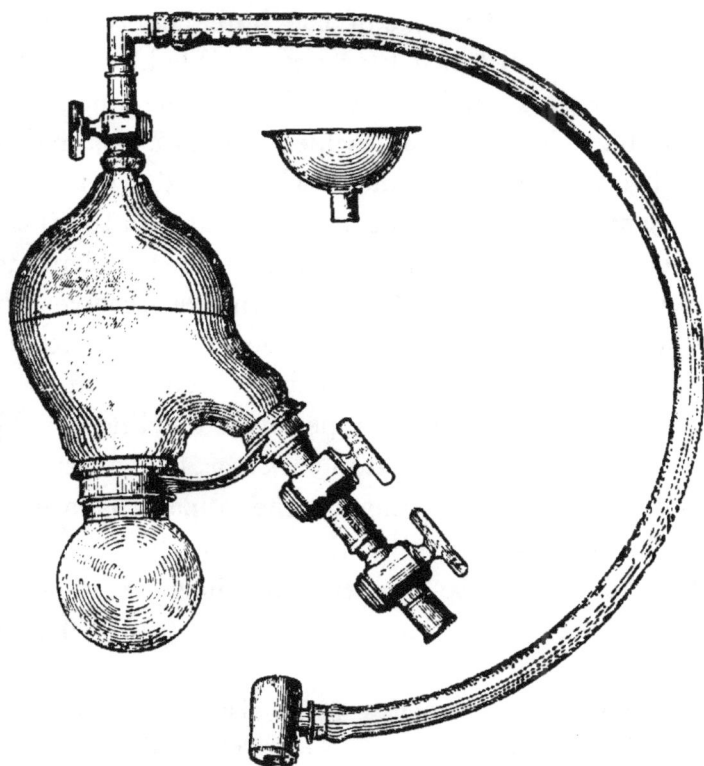

FIG. 8.

cock close to its junction with the bulb. By this hose water can be introduced into the aspirator from a neighbouring vessel, without disturbing the apparatus. There is also an extra stop-cock for the evacuating

C

catheter, the straight form of which is represented in the lower part of the woodcut.

In Bigelow's most recent aspirator * (Fig. 8), the long flexible elastic tube intervening between the bulb and the evacuating catheter is dispensed with, the catheter fitting into a brass tube, provided with a tap, inserted into the side of the bulb near the glass-receiver. The distance between the bladder and the aspirator is thus much shortened. At the upper part of the bulb is a tap, by which air escapes, and is excluded from the apparatus when filled with water. An essential feature of the new aspirator is the intro-duction of a strainer (not shown in the woodcut) for preventing the return of débris from the receiver into the bladder. This strainer is formed by a prolongation within the bulb of the brass tube, which fits on to the catheter, in the form of a perforated cylinder. The hose, funnel, and extra stop-cock for the evacuating catheter, though belonging to the apparatus, are not essential to it, and may be dispensed with.

We now come to the modifications of Bigelow's aspirator used by Sir Henry Thompson, of which there are three.

The earliest variety (Fig. 9) consists of a stout india-rubber bottle (A), on the upper part of which is a tap (B), and above this a small funnel through which the bottle is to be filled. At the lower end is a brass tube, attached to which are the lower tap (C) and the glass-receiver (E). The evacuating catheter is applied

---

* *Lancet*, Jan. 13, 1883.

at (D).   The mechanism of the lower tap is shown in
the woodcut.   The brass tube, to which the glass-
receiver is attached by a bayonet joint, or screw, is
continued down a short way into the glass globe, thus

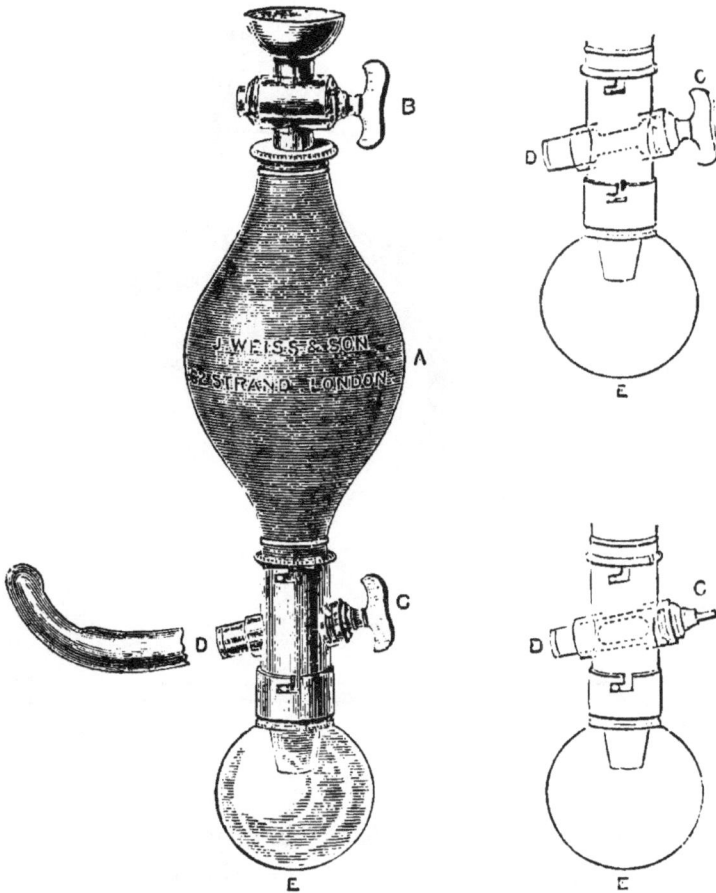

FIG. 9.

acting as a trap to prevent the return of the fragments
from the receiver.

In the aspirator just described, the glass-receiver is

c 2

placed directly beneath the india-rubber bulb, and it
is found in practice that the débris in the receiver is
disturbed by the currents of water produced by the
alternate compression and expansion of the bulb.

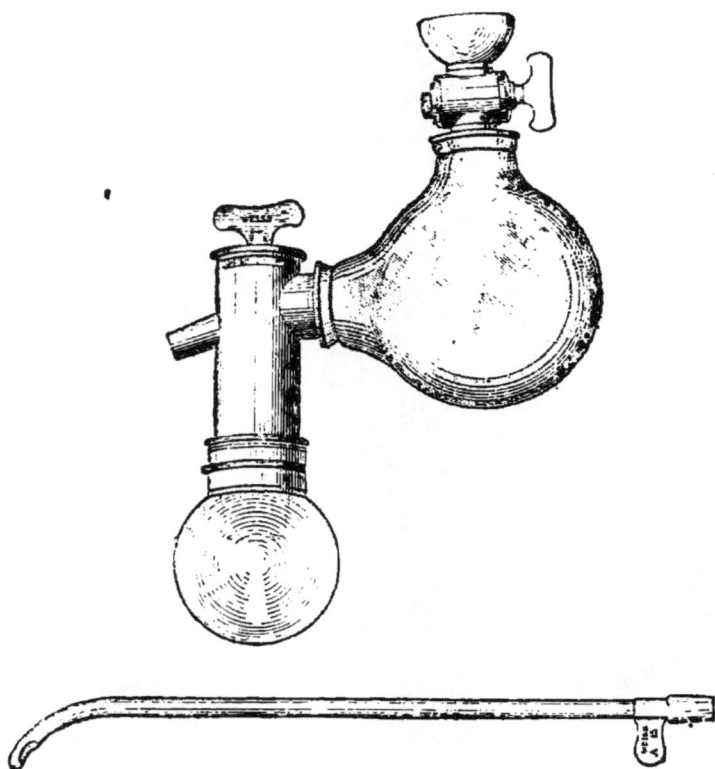

Fig. 10.

Some of the débris passes back into the bulb, and from
there a portion is carried again into the bladder by the
return stream.

To obviate this, Messrs. Weiss & Son suggested
that the brass cylinder with glass receiver be removed
to the front of the india-rubber bulb (Fig. 10), in

which position its contents would be less influenced
by the currents passing over the mouth of the receiver.
When the aspirator is in action, the greater portion of
the débris falls down into the receiver as the stream
from the bladder, diminished in force, passes the
empty chamber over its mouth. Some, however,
enters the bulb, and is carried back again with the
reverse stream; but, the catheter opening in the
cylindrical chamber being much smaller than the
opening into the bulb, most of the fragments impinge
against the sides and front of the cylindrical chamber,
and fall down into the receiver. Still a little débris
does pass back into the bladder, to be again withdrawn. Now, Bigelow remarks :*—" As we may
fairly assume that a surgeon would not deliberately
inject foreign bodies into a patient's bladder, there
must be something wrong in a system which obliges
him to do this, and makes it necessary to aspirate the
same débris twenty times over in order to remove it.
In short, the apparatus, as commonly arranged, is still
a defective one, and needs some special contrivance to
assist the action of gravity in securing the débris."
Admitting the force of these remarks, the objection
raised is, to a certain extent, theoretical. I have
worked with this variety of Sir Henry Thompson's
aspirator more than with any other, and have always
found it thoroughly efficient.

In Fig. 11 is illustrated Sir Henry Thompson's
most recent form of aspirator.† For the brass cylin-

---

* *Lancet*, Jan. 6, 1883, p. 6.   † *Ibid*. April 12, 1884, p. 653.

drical chamber and globular glass-receiver in his previous instruments we have now substituted a plain
glass cylindrical receiver, somewhat resembling the

Fig. 11.

glass trap in Clover's original syringe. A special
feature in this aspirator is the introduction of a light

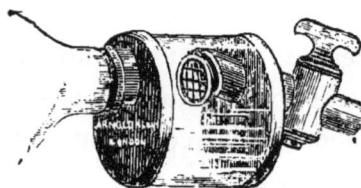

Fig. 12.

wire valve, attached inside
the chamber for the débris,
to the tube which receives
the evacuating catheter.
In Fig. 12 is shown a
diagonal view of this valve.
Its action is thus described
by Sir Henry :—" When pressure is made on the india-
rubber globe, and the current flows by the evacuating
catheter into the bladder, this light valve is driven close
to the aperture, and no débris can leave the glass trap.
When pressure is removed, and the current returns
from the bladder, the valve floats widely open, and
permits the débris to enter unchecked."

This recent form of aspirator, which Sir Henry Thompson says* is "not far removed from the original pattern of Clover," is not nearly so efficient an evacuator as his previous one (Fig. 10). The current of water passes directly into the receiver and disturbs the débris lying there, and the valve, as Sir Henry himself admits, "is sometimes liable to be partially blocked, as when mucus and fine débris are present." In fact, a previous remark† of Sir Henry Thompson, when criticizing Bigelow's views as to the construction of a new aspirator,—"All the perforated tubes and strainers get so blocked with débris (as I found long since) in the human body—not with coal in water— as to be practically useless there,"—is, to a large extent, applicable to his new form of aspirator. In the desire to produce an aspirator resembling the original syringe of Clover, efficiency is sacrificed, for the sake of assigning to the latter apparatus a position which neither the inventor nor any other surgeon, except Sir Henry Thompson, has claimed for it.

An aspirator designed by Mr. J. H. Morgan,‡ of Charing Cross Hospital, is illustrated at Fig. 13. I have never worked with it ; but Mr. Reginald Harrison, of Liverpool, and others report favourably on it. It seems to be light and handy, but it has the defect of having to be filled under water, and has no tap at the upper part for air to escape by.

Such, then, are the chief varieties of aspirator before the profession. There are many others, which it will

* *Lancet*, April 12, 1884, p. 653.   † *Ibid.* vol. i. 1883.
‡ *Ibid.* Sept. 2, 1882, p. 349.

be unnecessary to refer to here.   In fact, there are few surgeons of repute who practise litholapaxy who have not designed aspirators to suit their own peculiar views. But they are, one and all, modified imitations of Bigelow's original instrument.   All those described are efficient evacuators ; but the most convenient and

Fig. 13.

efficient are Bigelow's simplified evacuator (Fig. 8) and Thompson's aspirator modified by Weiss (Fig. 10). Both these are excellent aspirators, and there is little to choose between them.   The simplified form of strainer introduced into the former renders it perhaps the most perfect instrument of the two. I must confess, however, a great liking for Thompson's instrument. One of this variety, made for me by Messrs. Weiss & Son, I have worked with for four years ; and though it has assisted at 130 operations, and been through three hot weathers in the plains of India, it is still as efficient as much newer instruments I possess-- a fact which speaks well for the india-rubber employed in its construction.

THE OPERATION OF LITHOLAPAXY.

HAVING in the previous section given a sketch of the special instruments employed in the operation of litholapaxy, I will now proceed to describe in detail the various stages of the operation. I will assume that a stone of moderate size has been diagnosed in an adult, and its presence confirmed by the sound.

It will be well in all cases to submit the patient to preliminary treatment for a few days previous to undertaking the operation. This is especially necessary in a country like India, where patients frequently travel dozens, sometimes hundreds, of miles to some particular hospital reputed for the treatment of stone. The patient should be put to bed and placed on a light, nourishing diet. The bowels should be regulated, a purgative, such as castor oil, being given should constipation exist, and mild astringents should the patient suffer from diarrhœa,—which is often the case in this country. Barley-water, with alkalies and tincture of hyoscyamus, should be given if there be much irritation of the bladder present.

But the surgeon practising in India will say : " It is all very well to talk of keeping patients under preliminary treatment for some days previous to operation ;

this will often be impossible." This is very true. We all know the peculiar temperament of natives. They will come hundreds of miles to have an operation performed, and eagerly clamour to get on to the operating table, but often change their minds should there be any delay in its performance. They cannot understand the necessity of delay for preparatory treatment. We shall, therefore, have to take them when they are in the humour, and operate at once, should it be ascertained that they are likely to abscond. It has frequently happened to me, as it has to others in India, that a patient suffering from stone has come into hospital during my morning visit, has been placed on the operating table at once, and the operation there and then performed. This is one of the disadvantages under which surgeons in this country have to operate, and, in comparing the results obtained with those in European countries, should be taken into consideration.

For the operation, I find a low narrow operating table the most convenient. The patient is placed on this, close to the right edge, with his head resting on a pillow. The buttocks are raised by means of a low cushion placed beneath them. This is an important point, as the stone thus gravitates to the base of the bladder, away from the neck, and renders the latter part, which is the most sensitive, less likely to be injured in the various manipulations. The legs and thighs are flexed and slightly abducted, and held in this position by an assistant on either side.

In the cold weather, it is very essential that the patient should be protected by warm clothing during

the operation. For this purpose a pair of large
woollen stockings should be slipped on, reaching right
up the thighs close to the groins. Such a pair of
stockings can be made loosely out of an old blanket.
In addition to the ordinary clothing, a light blanket
should be thrown over the chest. These precautions
may be dispensed with in the hot weather in India ;
but in the cold season they are most essential to
prevent the patient from chill and consequent fever.

Close to the operator's right hand should be placed
a small stand, or teapoy, with a metal basin containing
warm water, in which the lithotrites and evacuating
catheters should be placed ready for use ; while the
aspirator, previously filled with warm water to the
temperature of the body, should be entrusted to an
assistant conversant with its working. It is always
well, when possible, to have two aspirators ready at
hand, to be used alternately. The operation is thus
facilitated, as, while the surgeon is using one, the
second can be emptied of débris, and refilled with
water by the assistant.

A small cup, or gallipot, containing oil, should be at
hand for lubricating the instruments. In this operation too much oil cannot be used, the instruments being
well oiled at each introduction.

Before undertaking the operation of litholapaxy the
surgeon must learn to pass all instruments—lithotrites,
sounds, and catheters—on the right side. This requires
only a little practice to do it with ease, and much time is
saved thereby. Besides the loss of time involved, it is
extremely awkward to see a surgeon passing the instru-

ments on the left side, and then going round to the right side to use them.

The patient now being anæsthetized, the surgeon, standing on his right side, should first pass a large conical steel sound into the bladder.

A series of sounds of this kind (Fig. 14), from No. 12 to No. 18, should be at hand. They should be made slightly tapering at the point, so that the diameter there is two sizes smaller than higher up. Solid heavy sounds of this kind are easily passed, and are handy for ascertaining the capacity of the urethra. and for facilitating the passage of other instruments. It will frequently be found that, when a catheter or lithotrite will not pass into the bladder, a heavy sound of this shape will; and on its withdrawal the lithotrite or canula may be passed.

The *meatus* is, as a rule, the narrowest part of the urethra, and it will frequently be found necessary, in order to pass the large instruments employed in litholapaxy, to enlarge it slightly. Should, therefore, the large sound not pass, this must be done at once. For this purpose a director is introduced into the urethra, and the floor of the *meatus* incised by means of a long slender scalpel. Or a urethrotome may be employed for this purpose. The operation is a very harmless one, and frequently procures an improvement on Nature.

FIG. 14.

The question now arises as to the quantity of water the bladder should contain during the crushing of the stone. As a rule, a very small quantity, from one to three or four ounces, will be sufficient to protect the walls of the bladder and, at the same time, permit of the necessary movements of the lithotrite. A large quantity of water is objectionable, involving an increased area over which the fragments, impelled by the currents set up by the movements of the lithotrite, may roam ; and thus increasing the difficulty in catching them. If, on the other hand, the bladder be completely empty, injury to its walls may result from the lithotrite. For my own part, I am indifferent as to the quantity of water the bladder may contain, provided it be not too large.

The lithotrite is now introduced. The operator stands obliquely with his left side towards the patient's face. The lithotrite, previously screwed home, locked and oiled, is held in the right hand by the cylindrical handle with the beak pointing downwards. The penis is grasped between the thumb and two first fingers of the left hand, and the beak of the instrument introduced into the urethra, the penis being drawn slowly but steadily on to the lithotrite, which is gradually elevated till it reaches the perpendicular position as it slides along the canal, which it does by its own weight. The beak will now have entered the membranous portion of the urethra as it passes through the triangular ligament. By gently depressing the handle of the lithotrite in the middle line towards the horizontal position, the beak will be found to slip along the membranous and prostatic portions of the urethra and into the bladder. As Sir Henry Thompson truly remarks:—" There is no more easy instrument to pass than the lithotrite with proper management."

The lithotrite being thus introduced, the next stage of the proceedings consists in catching the stone. For this purpose the lithotrite is passed gently onwards, or rather allowed to proceed by its own weight, along the trigone, till it reaches the most dependent part of the base of the bladder, on which it is allowed to rest. The instrument is then unlocked and the blades opened by withdrawing the male blade an inch or more, according to the size of the stone : the female blade being held steadily in position by the left hand on the cylindrical handle. The blades are now closed, when, frequently, the stone will be found between them. The lithotrite is locked and lifted slightly off the base of the bladder, and the stone crushed by screwing the male blade home. The instrument is again unlocked, the blades opened and closed, when a fragment will be caught, and crushed as before. This process is to be repeated several times, till a considerable quantity of fine débris is made. Our great master of the lithotrite—Sir Henry Thompson—compares the finding of fragments to fishing for perch : where one is found there will many be caught. We must not go searching about the bladder for fragments till those in the locality in which the stone is first found are disposed of. The depression in the base of the bladder, caused by the weight of the lithotrite resting on it, facilitates the stone, and subsequently its fragments, falling on to the female blade. A distinguished member of the Indian Medical Service, before whom I had the pleasure of operating, writing a few days subsequently of the features in the operation that astonished him, says :—"The next was the apparent ease with which the stone first, and afterwards its fragments, tumbled into the jaws of the instrument. It almost seemed as though they were anxious to get crushed."

Should the stone not be found by the manœuvres above indicated, it must be searched for. This is done by opening the blades of the lithotrite, turning them at an angle of 45° towards the right, and again towards the left, and closing them in these positions respectively. Should the stone still evade detection, the handle of the lithotrite must be depressed towards the horizontal position between the thighs, pushed an inch or so towards the fundus, and the same manœuvres gone through in that position, searching centrally, right and left. The stone will probably be found in one of these positions : but sometimes it lies immediately behind the prostate, especially when that gland is enlarged. To grasp the stone in this position, the handle of the lithotrite should be depressed between the thighs, and turned right round on its axis, so that the beak points downwards towards the trigone, but should not touch it. The blades are then opened and closed as before in this position, and if the stone lies there it will be secured. In fact, the surgeon should make a mental survey of the whole bladder, and institute a methodical search of every part of it, till the calculus is found. All the movements must be light and graceful, and care taken that the mucous membrane is neither caught between the blades nor otherwise injured. In whatever position found, the stone must be brought to the centre of the bladder, and there disposed of.

Let us now assume that the stone, or a portion of it, if a large one, has been reduced to fine débris. Should the stone be a small one—say, from a few grains up to 3 or 4 drachms in weight—its complete pulverization will probably be accomplished before the lithotrite is withdrawn, in a period varying from one to eight or ten minutes. But should the stone be a large one, a considerable amount of crushing, lasting over ten minutes or so, must be effected before removing the instrument. Before withdrawing the lithotrite, it must be locked and the blades screwed tightly home, so as to render them free of débris. I may here say that no instrument should be withdrawn from the bladder till quite free of débris.

The evacuating catheter should now be passed into the bladder, the largest size that the capacity of the urethra will admit being used. A rush of water and débris will take place, to receive which a small tray or bleeding basin should be at hand. The experience already gained in passing the solid sounds and lithotrites will afford a rough estimate of the size of canula that the urethra will admit. Should a No. 13 pass with ease, a No. 15 or 16 may be employed ; and should this latter be introduced with facility, a No. 17 or 18 may be tried. As to the shape of the canula, I prefer, as already stated, the curved variety ; but there is, as a rule, little difficulty in introducing either.

D

The canula having been introduced into the bladder, the aspirator, previously filled with warm water, is applied, the tap turned on, and aspiration of the débris begun. The right hand grasps the bulb of the aspirator, by the compression and expansion of which water is injected into, and withdrawn from, the bladder. With the outward stream the fragments are carried, and are seen to fall down into the glass-receiver, where they are trapped. Should the stone be a small one, and have been completely crushed at the first introduction of the lithotrite, it will be found that, after the aspiration has gone on for a time, the whole of the débris will have passed into the receiver. But if the stone be a large one, after a considerable quantity of débris has entered the receiver, which will vary with the amount of crushing at the first introduction of the lithotrite, little or no débris returns with the outward stream, but a rattling sound takes place, due to the fragments too large to pass being carried with force against the eye of the canula.

The aspirator is then removed, the canula withdrawn, and the lithotrite again introduced for the purpose of crushing more fragments. This is followed by the aspirator and canula as before. This process may have to be repeated several times, according to the size of the stone, before the whole of the débris is removed. I have removed successfully several calculi of two and three ounces in weight, requiring ten or twelve introductions of the instruments.

Such, then, is a general description of the operation. There are, however, difficulties met with and points to be attended to, to which I wish to draw attention here.

In a healthy urethra there are only two situations, as a rule, where difficulty may be encountered in the passage of instruments—viz., the triangular ligament and the neck of the bladder. The instrument (lithotrite or canula) should first be passed as far as it will go in the direction of the anus, thus depressing the floor of the urethra in front of the triangular ligament. " Traction on the penis next effaces this depression, and adds firmness to the urethral walls ; so that, if the instrument be withdrawn a little, and, at the same time, guided by the bony arch above, it can be coaxed without difficulty through the ligament in question— a natural obstruction which physicians often mistake for a stricture. The obstruction passed, the rest of the canal is short, and corresponds with the axis of the body " (Bigelow).

The obstruction sometimes met with at the neck of the bladder is due to the firm lower edge of the inner meatus. This may be overcome by pushing the lithotrite or canula gently onwards in the direction of the axis of the body, imparting to it a slightly rotatory motion, if necessary.

During the earlier part of the process of aspiration, the end of the canula should be kept towards the centre of the bladder, raised from the base, and may be moved about slightly in various directions to facilitate the flow of the fragments towards the eye ; but towards the completion of the process, the canula should be allowed to rest on the base, so as to gather up the sand and last fragments.

Towards the completion of the operation it will be

found, as a rule, that the last particles of débris lie close to the neck of the bladder, just behind the prostate. This is due to the fact that, the eye of the canula being turned towards the fundus and sides of the bladder, the water is less disturbed by currents in the position referred to than in any other. Consequently, the last particles of débris gravitate towards the spot. Towards the end of the operation, therefore, the eye of the canula should always be turned right round towards the prostate, and water forcibly injected, so as to dislodge the débris from this position. This manœuvre is especially necessary where enlargement of the prostate co-exists ; otherwise a fragment might be left behind.

On compressing the bulb and pumping water into the bladder, the débris is scattered away from the eye of the canula. Before allowing the stream to return by the expansion of the bulb, the hand should rest a few seconds, so as to allow the débris to settle down again in the vicinity of the eye. The evacuation of the débris will sometimes be found to take place best by injecting three or four ounces of water into the bladder with each compression of the bulb. Sometimes a much smaller quantity will be found most effectual. No definite rule can be laid down for all cases.

Sir Henry Thompson lays stress on the necessity of having the movements of the aspirator synchronous with those of the chest during respiration—the water being pumped into the bladder during expiration, and exhausted therefrom during inspiration. During my earlier operations I had recourse to this manœuvre a

good deal; but I came long ago to regard it as an unnecessary, and, indeed, frequently impracticable, refine-
ment.   Patients vary so much in the rapidity of their
breathing under the influence of an anæsthetic that the
suggestion frequently cannot be attended to, even if
desired.

It sometimes happens, even when the patient is fully
anæsthetized, that spasm of the bladder occurs.   During
its existence all manipulation should be suspended ;
otherwise the bladder might be injured.   This is a
point to which Mr. Reginald Harrison has called attention,* and the precaution indicated is a wise one.
Should the lithotrite be in the bladder, it must be closed
and kept unmoved till the spasm passes over.   If the
canula be in the bladder, the water should be allowed
to escape.

During the process of aspiration, with each expansion of the india-rubber bulb, the fragments of calculi
are carried against the eye of the canula by the outward
rush of water, and a clicking sound is thus produced,
which, whilst it continues, indicates that some fragments remain in the bladder.   There is, however,
a peculiar sound sometimes produced, which I have not
seen mentioned in any text-book or journal, and the
occurrence of which the young litholapaxist should be
acquainted with, as it is very likely to be confounded
with the sound produced by a fragment.   This " false
sound," as it may be called, is produced by the mucous
membrane of the bladder being sucked into the eye of

---

* "Lithotomy, Lithotrity, &c.," by R. Harrison, 1883, p. 35.

the canula during exhaustion of the water. It is most likely to occur towards the end of the operation, when all, or nearly all, the fragments have been exhausted, and especially when the bladder contains no surplus water, only that quantity which is pumped in and withdrawn during compression and expansion, respectively, of the bulb. It may, however, be produced at any time if, after compressing the bulb, the eye of the canula be turned towards the sides, or directed up against the fundus of the bladder, and then the bulb of the aspirator be allowed to expand. The sound itself, though difficult to describe, can never be mistaken when once recognized. The sensation communicated to the hand is of a fluttering, jerky character, accompanied by a dull, muffled sound as contrasted with the clear, ringing click which the impact of fragments imparts to the instrument. On its occurrence the outward stream receives a sudden and complete check; whereas, when a fragment obstructs the stream, a portion of the water continues to flow. The sound does not recur if the canula be partially withdrawn and raised towards the perpendicular position, so as to bring the eye close to the neck of the bladder, with the end of the canula resting on the trigone; whereas a fragment will produce obstruction there, as well as in any other position. On first practising litholapaxy I was deceived by this sound, and recently I have seen a brother-officer of the Indian Medical Service, who performed his first litholapaxy operation in my presence, similarly deceived, I having forgotten to mention to him the possibility of its occurrence.

It frequently happens that, during the process of aspiration, a fragment which is too large to pass through the canula gets caught in its eye. This is recognized by the fact that the outward stream is arrested, and the bulb of the aspirator ceases to expand. The fragment should at once be displaced. This, as a rule, may be effected by compressing the bulb suddenly and with force, when the fragment will be expelled by the inward stream. Should this manœuvre fail after being tried two or three times, a stylet, or gum-elastic bougie, should be introduced through the canula, and the fragment displaced in this way. But the canula should on no account be withdrawn with the fragment sticking in its eye, as in this way the urethra may be injured, or the fragment get caught in the mucous membrane, and thus impacted in the urethra.

Should a fragment get impacted in the urethra, how are we to deal with it ? This is an accident that has never occurred to me ; and, if the precautions indicated in the last paragraph are taken, there will be little fear of its occurrence. Still, it is an accident that has to be reckoned on. Should it occur, the fragment may be removed by long urethral forceps (Figs. 15, 16, 17).

Where there is great irregularity of the inner surface of the bladder, it may be extremely difficult to get rid of the last fragment. I have experienced this more than once. The aspirator is applied, and time after time the fragment clicks against the eye of the canula, but, on introduction of the lithotrite, the fragment cannot be grasped. Now it is in such cases as this that a shallow, flat-bladed lithotrite is of most use.

Great perseverance may be necessary, especially if the fragment be a broad, thin shell from the outer crust of a large stone.   A manœuvre that I have found useful is to employ the suction force of the canula and aspi-

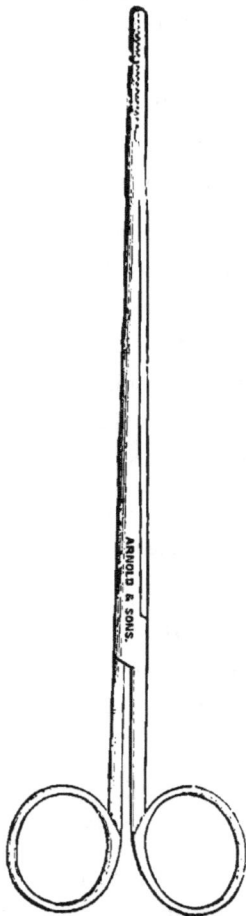

FIG. 15.   FIG. 16.   FIG. 17.

rator to bring the fragment close to the neck of the bladder, and then to introduce the lithotrite and catch the fragment in this position.

The larger the evacuating canula, the less necessity there will be for crushing the calculus into fine powder, and, consequently, the less time will the operation require for its performance—a matter of no small importance when we have to deal with a large stone in a patient whose constitution has been very much worn by the disease. It is, therefore, advisable to employ the largest canula that will pass with ease into the bladder. I cannot too strongly deprecate the use of any force in passing a catheter, or, indeed, any instrument, into the bladder; but the deleterious effects which Sir Henry Thompson anticipated from the use of large instruments, experience has shown to be mythical. Sir Henry says that the instruments should be proportionate to the size of the stone; but experience has taught me that the capacity of the urethral canal is of much more importance in determining the size of the instruments, and that the largest lithotrite and canula that can be passed without the use of any force should be employed. A large lithotrite is much handier in the bladder, less liable to get clogged by débris, much more efficient, not only for crushing large calculi, but for disposing of fragments of débris, than a small one; and I fully agree with Bigelow that, when one gets accustomed to the use of a large lithotrite, he does not willingly abandon it for a smaller instrument.

It will be absolutely necessary to have recourse to large instruments much more frequently in this country than in England, for the simple reason that the great majority of calculi are large when coming under observation in India.

And here I must refer to a most unaccountable statement that occurs in the latest edition of Bryant's Surgery. In describing Bigelow's operation, the following
passage occurs :*—" A No. 15 or 16 evacuating catheter
for small, and a No. 17 or 18 for large, stones should be
selected and passed, *the larger instruments, Nos.* 25 *to*
30, *as used by Bigelow, being required in exceptional cases
only* [the italics are mine], the size being determined
by that of the urethra." Now, no such instruments as
those ascribed to Bigelow in the passage in italics have
ever been used by him, so far as I am aware. The
largest catheter recommended in his writings is a No.
20, English, and this for use in exceptional cases only,
Nos. 17 and 18 being those most relied on for general
use. It is extremely to be regretted that an error of
this kind should have crept into what is, perhaps, the
most popular text-book of surgery in the English lan-
guage. A statement of this kind is calculated, not
only to misrepresent the distinguished author of the
operation, but to prejudice students against the
operation.

When I first read this passage, I was very forcibly
reminded of an incident of rather an amusing character
that once occurred to me. A Hospital Assistant, who
had come a long distance to witness the operation of
litholapaxy, was extremely astonished on seeing the
instruments displayed on a table before him. Taking
up the No. 20 canula, he naïvely inquired whether in
use it was placed *in*-side or *out*-side the penis ! The

---

* "Practice of Surgery," by T. Bryant, fourth edition, vol. ii. p. 121.

question was, after all, not a very unnatural one for a
man who had never previously seen a larger catheter
than No. 12.   Certainly, had he seen the imaginary
canulæ Nos. 25 to 30 above ascribed to Bigelow the
remark would have been a very natural one.

In the whole history of surgery I suppose there has
been no operation, involving, as this does, one of the
greatest improvements ever introduced into operative
surgery, that has met with so much adverse criticism
and misrepresentation as Bigelow's.   Out here in India
I have seen instruments which are a gross libel on the
originator of the operation.   Some instrument-makers
are in the habit of sending out the apparatus with
canulæ Nos. 18 and 20 only, as if catheters of this size
alone were used ; the fact being that they were recommended in exceptional cases only.

As a rule, there is little or no loss of blood attending
the operation, with the exception of the trifling bleeding that follows the incision in the floor of the urethra
where this is necessary to enlarge the *meatus*.   I have
frequently removed very large calculi with scarcely a
tinge of blood in the washings from beginning to end.
In some cases, however, the mucous membrane of the
urethra is highly sensitive to the passage of instruments.
and considerable bleeding takes place.   In such cases
I am in the habit of using a weak astringent in the
washings, say, ¼ grain of acetate of lead to the ounce,
and winding up the proceedings with a stronger solution
(one grain to the ounce).

The operation being now completed, the patient
should be put to bed, and well wrapt up in warm

clothing. A morphia suppository should be at once introduced. As soon as the patient recovers consciousness, I am in the habit of administering a large dose of quinine. The quinine may be repeated in smaller doses for a few days. There is no doubt that the administration of this drug has the effect of warding off the fever, which, in this country, is especially liable to supervene on any operation of this nature. The food for the first few days should be of a light kind, consisting mainly of milk and soups. A demulcent and alkaline drink should be allowed. My favourite mixture is a quart of barley-water, mixed with which are one drachm each of liquor potassæ and tinct. hyoscyami ; and this the patient is encouraged to drink freely.

For the first twenty or thirty hours the urine will be tinged with blood ; and there will, as a rule, be considerable burning sensation along the course of the urethra, with some difficulty of micturition. The treatment indicated in the last paragraph will tend to alleviate these symptoms. Should there be any pain, or tenderness on pressure, in the region of the bladder, hot fomentations assiduously applied. followed by hot poultices to the hypogastric region, will be found soothing ; and pain in the perineal region will be lessened by painting the part with extract of belladonna.

Retention of urine is a rare sequel of the operation, for which a hot hip bath will be found most effectual. Should this fail, recourse must be had to the catheter. More rare still is total suppression of urine, which should be dealt with on general medical principles.

When the patient is the subject of atony of the
bladder or enlargement of the prostate, it may be
advisable to pass and tie in a soft gum-elastic catheter
for a few days, to allow the water to flow in this way.
When stricture of the urethra exists, and has been
the subject of treatment before the operation, it will
always be advisable to have recourse to this precaution.

Acute inflammation of the testicle is a sequel of the
operation that has twice occurred in my practice, and
readily yielded to the ordinary treatment for that
complication.

But the most frequent sequel of the operation in
this country is the occurrence of fever.   To the ordinary
catheter or urethral fever, long recognized as attending
the passage of instruments through the urethra, we
have superadded, as it were, the effects of malaria in
India ; and the supervention of this fever is a con-
tingency that we shall have to reckon with in a large
proportion of the cases dealt with.   The attack sets in,
as a rule, two or three hours after the operation, and
passes through the usual stages—*cold, hot,* and *sweating*
—of an ordinary attack of intermittent fever, from
which it is scarcely to be distinguished.   The treatment will also be the same as in ague,—extra warm
clothing, hot-water bottles to the extremities. and the
administration of hot drinks, particularly tea, during
the cold stage.   As this passes into the hot stage, part
of the clothing must be removed, and the patient's
thirst relieved by copious drinks of water, lemonade,
&c.   The ordinary fever mixture should also be given
to encourage perspiration.   When the sweating stage

sets in, warm clothing must be again supplied to encourage perspiration and prevent the patient catching cold. During the intermission, quinine should be given. The fever is, as a rule, very amenable to treatment.

In the preceding pages I have endeavoured to give as clear a description as possible of the instruments employed, and the various details of the operation of litholapaxy. The operation is, undoubtedly, a difficult one, perhaps the most difficult in the whole range of operative surgery, and should not be lightly undertaken by inexperienced hands. There are some men whose hands were never made for the use of surgical instruments, and my advice to such is, not to undertake the operation. There are other spheres in the medical profession of as much usefulness as that of operative surgery. In 1867, Sir William Fergusson, writing of lithotrity, says : " I know not any process in surgery requiring more forethought, knowledge, manipulative skill, and after-judgment." And if this remark of one of the most distinguished lithotritists of his day was true of the old operation of lithotrity, how much more is it applicable to the modern operation of litholapaxy, in which instruments of much larger size and greater power than formerly used are employed, in which calculi of much larger dimensions are attacked, and in which the proceedings are extended over much longer periods. Patience, perseverance, gentleness, dexterity, a light touch, and, above all, experience are essential to make a man a good litholapaxist. I do not hesitate to say that, on every occasion that I have performed the operation, I have learnt something new.

In this fact consists one of the great beauties of the operation, so far as the surgeon is concerned. It is always a field of novel research.

It has been said that no novice should undertake this operation; and this is undoubtedly true so far as the general practitioner in England is concerned, who passes months, sometimes years, without undertaking an operation of any magnitude. But the case is different in India. There are few officers of the Indian Medical Service, holding civil appointments, who do not reckon their major operations annually by the score, frequently by the hundred, and in some instances by the thousand. During the past few years a healthy spirit of rivalry, as to the amount of good work done, has entered the ranks of the service; and I do not hesitate to say that there are now many hospitals in the North-Western Provinces, under the Government of which I have the honour to serve, where more operations are undertaken single-handed by the Civil Surgeon than by the whole staff in some of the largest London hospitals. It goes without saying that the large experience thus acquired must give to the surgeon practising in India a manipulative dexterity in the use of surgical instruments, an amount of self-reliance and judgment in dealing with difficulties that unexpectedly arise, which can be rarely acquired elsewhere. This being the case, why should he hesitate to undertake the operation of litholapaxy?

The beginner will do well to commence by operating on cases where the stone is small and the urethra capacious. As experience is acquired, large calculi and those attended by complications are to be attacked.

Previous to undertaking the operation for the first time, it will be always well, when possible, for the surgeon to pay a visit to one of those hospitals where the operation is performed. More information will be gained by seeing the operation once well performed than from any amount of reading and theoretical knowledge. It is not the object of this work to take the place of practical knowledge, but rather to assist and supplement it.

THE AUTHOR'S EXPERIENCE OF
THE OPERATION,

WITH COGNATE STATISTICS.

In the preceding pages I have endeavoured to give as clear a description as possible of the operation of litholapaxy and of the instruments employed in its performance. I will now give the results of my own experience of the operation, with some cognate statistics.

During the period that has elapsed since I adopted the operation of litholapaxy in my practice, 225 cases of stone in the urinary passages have come under my immediate treatment. Amongst these there were 4 cases of impacted urethral calculus, in all of which external urethrotomy was successfully performed. There were 79 cases of vesical calculus in male children or lads under the age of 16. Lateral lithotomy was the operation performed by me in all these cases, and amongst them there was no death, all having made excellent recoveries. In the remaining 142 cases, all of which occurred in adult males, with the exception of 4 females, the operation of litholapaxy was entertained, but for various reasons there

E

were 14 instances in which the operation could not be performed. Five of these latter were the subjects of severe stricture of the urethra, so that the instruments would not pass into the bladder; in 2 cases there was a greatly enlarged prostate with tortuous urethral canal; and in 7 cases the stone was so large that it could not be grasped by the lithotrite, or so hard that, after being grasped, the lithotrite could make no impression on it. In 13 of these cases I performed lateral lithotomy, with 2 deaths. In the remaining case the calculus was extremely large and hard, weighing over 12 ounces. In this case the supra-pubic operation was performed, but the patient died six hours afterwards.

I have considered it advisable to mention the above facts so that it may be clearly understood that the operation was given a fair trial, having been performed in every possible instance in the adult without reference to age, state of health, size of calculus, &c. It is in this way only that a true estimate of the comparative value of the operation can be obtained, not by performing it in selected cases only.

I may further add that during the period referred to no case of stone coming under my observation has been refused surgical relief. Every case of stone that came under observation during that period was operated on, with the exception of three, as far as my recollection goes. One was an old man emaciated from diarrhoea. He was kept in the Bareilly Civil Hospital for about a fortnight for the purpose of improving his general health before subjecting him to litholapaxy. Growing tired of waiting, he absconded. The second was rather a surgical curiosity. It was that of a child who was being chloroformed with a view to my performing lithotomy. When almost fully anæsthetized, he passed urine spasmodically and with great force, and with it the calculus—which was small—rushed out suddenly. The third instance was one in private practice—that of a native gentleman from Agra. After placing him under the influence of chloroform, it was found that the stone was larger than anticipated, so that the largest lithotrite would not close on it. I proposed lithotomy to the patient, but, as he had come specially for the operation of litholapaxy, and, being a Government servant, had only a limited time at his disposal, he was then unable to submit to the operation.

The following table will show at a glance some of the most important features of interest connected with the 128 cases of litholapaxy which are the text of this monograph :—

Table Showing Particulars of 128 Litholapaxy Operations.

Serial No.	Date of Operation	Age (Years)	Sex	Caste	No. of Days in Hospital	No. of Sittings	Size of Canula (No.)	Weight of Calculus Oz.	Drs.	Grs.	Variety	Time Occupied by Operation (Minutes)	Duration of Disease	Result
1	July 3, 1882	50	M.	H.*	10	1	18		4		Phosph.	30	9 months	Successful
2	,, 25, ,,	85	M.	M.+	10	1	16		6	20	Uric	25	3 years	,,
3	Aug. 4, ,,	55	M.	M.	13	1	16		1	55	,,	10	9 months	,,
4	Sept. 6, ,,	65	M.	M.	10	1	18		3	40	Oxalate	25	5 years	,,
5	,, 14, ,,	65	M.	H.	6	1	16		6		,,	25	4 ,,	,,
6	,, 15, ,,	50	M.	H.	5	1	18	1	4		Phosph.	18	2 ,,	Died
7	,, 22, ,,	65	M.	H.		1	18		3		Uric	33	2 ,,	Successful
8	,, 23, ,,	40	M.	H.	7	1	16		1	30	,,	5	1 ,,	,,
9	,, 24, ,,	40	M.	H.	6	1	18		7		Phosph.	25	3 months	,,
10	,, 26, ,,	60	M.	H.	7	1	16		2		Oxalate	12	2 years	,,
11	Oct. 17, ,,	45	M.	H.	9	1	18		7	50	Uric	25	4 ,,	,,
12	,, 23, ,,	20	M.	H.	5	1	16		2	15	Oxalate	11	10 days	,,
13	Nov. 2, ,,	50	M.	M.	5	1	18				,,	6	12 years	,,
14	,, 17, ,,	35	M.	H.	8	1	16		5	25	Phosph.	29	2 ,,	,,
15	,, 28, ,,	40	M.	M.	10	1	18		5	35	Uric	23	3 ,,	,,
16	Dec. 6, ,,	60	M.	M.	10	1	16	3	2		Phosph.	66	1 ,,	,,
17	,, 6, ,,	36	M.	H.	7	1	14		2	5	Uric	8	12 ,,	,,
18	,, 19, ,,	35	M.	M.	7	1	16			45	Phosph.	5	1 month	,,
19	,, 27, ,,	40	M.	M.	12	1	16	1	2	10	Oxalate	35	6 ,,	,,
20	Jan. 1, 1883	18	M.	H.	12	1	16			50	Uric	2	3 years	,,
21	,, 3, ,,	55	M.	M.	13	1	14				,,	10	2 days	,,
22	,, 6, ,,	41	M.	H.	5	1	18		5	15	Oxalate	25	9 months	,,
23	,, 9, ,,	48	M.	M.		1	18			20	Uric	5	2 years	Died
24	,, 17, ,,	32	M.	M.	11	1	18		1		Oxalate	8	2 ,,	,,
25	Feb. 17, ,,	17	M.	H.	9	1	16		5	40	Uric	35	2 ,,	Successful

No.	Date	Sex	Caste	Age					Composition		Duration	Result
27	" 19,	M.	H.	40	11	1	16			Phosph.	55	"
28	" 22,	M.	M.	40	7	1	18			Uric		"
29	March 5,	M.	H.	46	22	1	16			Phosph.	10	"
30	" 19,	M.	M.	35	8	1	16			"	5	"
31	April 4,	M.	H.	32	8	1	16			Uric	30	"
32	" 12,	M.	M.	35	8	1	18			Oxalate	35	"
33	" 12,	M.	H.	50	11	2	16			Uric		"
34	" 15,	M.	M.	85	4	1	16			Phosph.	40	"
35	" 16,	M.	H.	55	12	1	16			Uric	40	"
36	" 20,	M.	M.	65	13	1	16			"	20	"
37	" 21,	M.	H.	40	11	1	18			Oxalate (Carb. lime)	20	"
38	" 24,	M.	M.	36	6	1	16			Uric	10	"
39	" 27,	M.	H.	52	7	1	14			"	40	"
40	" 28,	M.	M.	50	11	1	18			Oxalate	30	"
41	May 8,	M.	H.	16	6	1	16			Uric	15	"
42	" 18,	M.	M.	45	8		18			Phosph.	40	"
43	" 19,	M.	H.	17	12	1	16			Uric	40	"
44	" 25,	M.	M.	45	9	1	18			Oxalate	12	"
45	" 28,	M.	H.	19	9	1	16			Uric	30	"
46	June 3,	M.	M.	35	8	1	18			"		"
47	" 21,	M.	H.	65	6	1	16			Oxalate	40	"
48	" 24,	M.	M.	85	12	1	16			Uric	30	"
49	July 3,	M.	H.	36	10	1	14			"	6	"
50	" 8,	F.	M.	65	8	1	16			"	10	"
51	" 10,	M.	H.	65	3	1	18			"	26	"
52	" 27,	M.	M.	18	3	1	18			Phosph.	44	"
53	Aug. 1,	M.	H.	50	12	1	16			Uric	21	"
54	" 10,	M.	M.	4½	10	1	16			Phosph.	40	"
55	" 12,	M.	H.	85	19	1	16			Uric	19	"
56	" 14,	M.	M.	65	5	1						"
57	" 22,	M.	H.	30								"
58	" 23,	M.	M.	32								"

+ Mahomedan. # Hindoo.

Table Showing Particulars of 128 Litholapaxy Operations—(continued).

Serial No.	Date of Operation	Age (Years)	Sex	Caste	No. of Days in Hospital	No. of Sittings	Size of Canula (No.)	Weight of Calculus (Oz.)	(Drs.)	(Grs.)	Variety	Operation, by Occupied Time (Minutes)	Duration of Disease	Result
60	Aug. 24, 1883	60	M.	H.	7	1	18	—	4	30	Uric	30	2½ years	Successful
61	" 25, "	45	M.	H.	12	1	16	—	2	5	"	10	1½ "	"
62	Sept. 3, "	50	M.	H.	6	1	18	—	—	32	"	5	1	"
63	" 16, "	36	M.	M.	6	1	16	—	2	26	Phosph.	10	¼ "	"
64	" 16, "	42	F.	H.	6	1	18	—	3	26	Uric	20	13 "	"
65	" 20, "	3¾	M.	H.	3	1	14	—	1	30	Phosph.	15	1	"
66	Oct. 2, "	4	F.	M.	5	1	14	—	—	57	"	10	3	"
67	" 15, "	65	M.	H.	15	1	16	—	1	15	Uric	5	"	"
68	" 26, "	55	M.	H.	5	1	18	—	—	35	"	34	8 months	"
69	" 28, "	36	M.	M.	15	1	18	1	1	35	"	60	4 years	"
70	Nov. 21, "	40	M.	M.	17	1	16	2	—	55	"	90	4 "	"
71	Dec. 4, "	50	M.	H.	8	1	16	—	—	25	Phosph.	6	"	"
72	" 5, "	35	M.	M.	7	1	18	—	—	11	Uric	3	"	"
73	" 11, "	50	M.	M.	6	1	18	—	2	45	Phosph.	8	"	"
74	" 13, "	60	M.	H.	5	1	16	—	1	46	"	10	1	"
75	" 14, "	26	M.	M.	12	1	18	—	2	35	Uric	20	3 "	"
76	" 27, "	45	M.	M.	1	1	16	—	—	15	"	1	2 months	"
77	Jan. 5, 1884	40	M.	H.	2	1	16	—	3	13	"	17	2 "	"
78	" 24, "	18	M.	M.	6	1	16	—	—	10	"	35	9 years	"
79	March 2, "	75	M.	H.	14	1	16	1	—	—	"	29	3 "	"
80	" 11, "	45	M.	M.	6	1	16	—	2	30	"	5	3 "	"
81	" 15, "	70	M.	H.	14	1	16	—	6	30	"	20	3 months	"
82	" 16, "	60	M.	M.	5	1	18	—	5	30	"	25	1 year	"
83	" 16, "	58	M.	H.	8	1	16	—	2	48	Phosph.	10	2¼ "	"
84	" 29, "	60	M.	H.	9	1	16	1	1	—	"		2 "	"
85	April 15, "	75	M.	H.	10	1	16	—	—	40	Uric	54	3	"

No.	Date	Age	Sex	Nat.							Composition		Duration	Result
86	,, 16,	60	M.	H.	9	1	14	—	1	54	Phosph.	15	1	,,
87	,, 25,	60	M.	M.	6	1	18	—	1	3	} Uric and Phosph.	5	3 months	,,
88	May 23,	45	M.	H.	11	1	16	1	4	—	Oxalate	11	2 years	,,
89	,, 23,	35	M.	H.	8	1	16	—	1	10	Uric	7	1	,,
90	June 7,	16	M.	H.	6	1	16	—	3	15	Phosph.	10	2½	,,
91	,, 21,	35	M.	M.	9	1	18	2	—	—	} Uric and Oxalate	17	2¾	,,
92	,, 27,	55	M.	H.	19	1	16	1	4	—	Uric	28	3	,,
93	,, 29,	55	M.	H.	12	1	16	—	5	10	,,	17	2½	,,
94	July 11,	45	M.	H.	7	1	18	—	3	15	,,	11	3	Died
95	,, 16,	55	M.	M.	9	1	14	1	1	20	Oxalate	27	1½–4	Successful
96	Aug. 1,	40	M.	M.	4	1	16	—	—	20	} Uric and Oxalate	4	6	,,
97	,, 13,	26	M.	M.	—	1	18	3	—	30	Uric	52	4	,,
98	,, 16,	75	M.	H.	9	1	16	—	1	25	Phosph.	6	1 month	,,
99	,, 18,	34	M.	M.	2	1	14	—	—	5	,,	2	3 years	,,
100	,, 24,	20	M.	H.	5	1	18	—	7	—	Uric	18	15 days	,,
101	,, 25,	56	M.	H.	11	1	16	—	—	15	Phosph.	5	3 years	,,
102	,, 26,	35	M.	M.	7	1	16	—	5	30	Uric	36	9 months	,,
103	Sept. 6,	80	M.	H.	9	1	16	—	1	20	,,	15	,,	,,
104	,, 8,	56	M.	H.	5	1	16	—	—	12	,,	5	4	,,
105	,, 10,	50	M.	M.	6	1	14	—	—	15	,,	5	1 year	,,
106	,, 13,	36	M.	H.	14	1	16	—	4	—	Oxalate	10	4	,,
107	,, 24,	45	M.	H.	10	1	18	1	—	11	} Uric and Phosph.	35	10 days	,,
108	,, 29,	25	M.	M.	14	1	16	1	2	5	Oxalate	20	3 months	,,
109	Oct. 6,	35	M.	M.	7	1	11	—	—	8	Phosph.	2	2½ years	,,
110	,, 11,	25	M.	H.	5	1	16	—	2	5	Oxalate	11	1 month	,,
111	,, 26,	55	M.	H.	9	1	18	2	—	45	Uric	26	9 years	,,
112	,, 27,	60	M.	E.*	3	1	14	—	—	30	Oxalate	5	1 year	,,
113	Nov. 11,	55	M.	H.	3	1	16	—	6	16		15	7 months	,,
114	,, 21,	46	M.	H.	12	1	14	—	1	20		14		,,
115	,, 28,	38	M.	M.	6	1	14	—	—	16		6		,,

* European.

Table Showing Particulars of 128 Litholapaxy Operations—(continued).

Serial No.	Date of Operation.	Age.	Sex.	Caste.	No. of Days in Hospital.	No. of Sittings.	Size of Canula.	Weight of Calculus. Oz.	Drs.	Grs.	Variety.	Time Occupied by Operation.	Duration of Disease.	Result.
		Years.										Minutes.		
116	Dec. 2, 1884	26	M.	H.	6	1	18	—	7	18	Uric	20	3 years	Successful
117	,, 6, ,,	55	M.	H.	7	1	18	—	9	—	,,	33	1½ ,,	,,
118	,, 20, ,,	40	M.	H.	7	1	16	—	1	5	,,	10	1 ,,	,,
119	Jan. 15, 1885	40	F.	H.	2	1	14	—	1	30	,,	10	1 ,,	Died
120	,, 27, ,,	61	M.	H.	—	1	14	—	—	18	,,	15	2 ,,	Successful
121	Mar. 20, ,,	32	M.	H.	14	1	18	—	5	30	Phosph.	23	5 ,,	,,
122	May 20, ,,	62	M.	E.	4	1	14	3	2	5	Uric	15	2 ,,	,,
123	Aug. 30, ,,	60	M.	M.	24	1	18	—	2	30	,,	65	30 ,,	,,
124	Sept. 3, ,,	56	M.	H.	5	1	18	—	3	25	Phosph.	25	4 ,,	,,
125	,, 7, ,,	45	M.	H.	6	1	18	2	4	40	Uric	15	2¾ ,,	,,
126	,, 20, ,,	55	M.	M.	8	1	16	—	2	—	,,	47	2 ,,	,,
127	Oct. 21, ,,	27	M.	E.	2	1	14	—	1	10	,,	5	6 months	,,
128	,, 26, ,,	40	M.	H.	2	1	16	—	1	30	,,	12	1 year	,,

The large majority of these operations were performed in the civil hospitals with which I have been
connected during the past four or five years, chiefly at
Bareilly and Moradabad, a comparatively small proportion having occurred in private practice. Careful and
detailed notes of all the cases have been taken by my
assistant-surgeons and myself. I may also mention
that in the great majority of the cases I had the
pleasure of operating in the presence of one or more
medical officers.

The average number of days spent in hospital, or, in
the case of patients in private practice, confined to the
house, was about 8·5. And I may here mention that
my practice is, never to allow litholapaxy cases to leave
hospital till their cures are complete. Indeed, my
rule is to keep them in hospital a day or two longer
than absolutely necessary, so as to be on the safe side.
Contrast this period with the lengthened stay in
hospital that lithotomy involves !

The male patients operated on varied from puberty
to 96 years, the average age being 47½ years. One
of the females was an adult, aged 36 ; the other 3
were under 5 years.

The 128 operations were performed on 126 different
individuals, the disease having only twice recurred.
In one of these cases the patient, aged 65, was
suffering from enlargement of the prostate. Six
months after the first operation, in which 4 drachms
of a soft phosphatic calculus were removed, he returned,
and had a similar calculus, weighing 85 grains, removed. In old patients of this kind, with enlarged

prostate, the disease frequently recurs, owing to the fact that the bladder is never completely emptied of urine during micturition. In the other case several small calculi, varying in size from that of a pea to that of an almond, and weighing in all 1 ounce, were removed, and the patient left the hospital quite well. A month afterwards he returned with symptoms of stone. At first I imagined that a fragment had been left behind at the first operation, but, on applying the canula and aspirator, I removed 27 minute calculi, weighing ½ drachm in all. I then found that in the interval the patient had suffered from severe kidney colic, and that the minute calculi, each of which consisted of a uric-acid nucleus with phosphatic deposit, were the result of a shower of uric-acid particles from the kidney.

The calculi removed varied in weight from 5 grains to 3¼ ounces. There were 57 weighing ½ ounce and upwards; 30, 1 ounce and more; 9, 2 ounces and upwards; and 3, more than 3 ounces.

The period occupied by the operation varied from two or three minutes to one hour and a half—the time required for the removal of 10½ drachms of uric-acid calculus from a man aged 80, in which there were considerable difficulties met with. It will be noticed that three calculi, weighing over 3 ounces each, were removed in a period of about one hour in each instance.

From the preceding table it will be observed that amongst the 128 litholapaxy operations there were 5 deaths. One of these, as will subsequently appear, was

not justly attributable to the operation, the patient
having died from bronchitis, cold and neglect. Accept-
ing the 5 deaths in connection with the operation,
however, we find that, excluding the 3 operations
on female children, there were 125 operations on adults
with 5 deaths, or a mortality of 4 per cent. Compare
these results with the recognized mortality of 1 in 4, or
25 per cent., from lithotomy in the adult !

An impression seems to prevail, both in this country
and in England, to the effect that lithotomy amongst
the natives of India is a much more successful operation
than amongst Europeans. Let us examine the facts of
the case.

In the end of 1884, when preparing an article
on litholapaxy for publication, Surgeon-General W.
Walker was good enough to lend me the latest pub-
lished Medical Administration Reports of the various
Presidencies and Provinces in India, and from these
I compile the following

Table Showing the Results of Lithotomy Operations performed in Indian Hospitals in 1882.

Presidency or Province.	Remaining December 31, 1881.	Performed during 1882.	Total.	RESULT.					Percentage Fatal.	Proportion Fatal.
				Cured.	Relieved.	Otherwise.	Died.	Remaining.		
Bengal..............	6	181	187	148	5	5	19	10	10·5	1 in 9·5
N.-W. Provinces and Oudh	49	933	982	824	11	11	85	51	9·1	1 in 11
Punjab	50	949	999	768	25	28	127	51	13·4	1 in 7·4
Central Provinces	5	73	78	68	...	1	9	...	12·3	1 in 8
Bombay and Sindh	27	450	477	377	6	6	45	43	10·0	1 in 10
Madras	1	6	7	5	1	...	1	...	14·4	1 in 9
Total.........	138	2,592	2,730	2,190	48	51	286	155	11·0	1 in 9

According to the above table, the mortality occurring
actually in hospital amongst 2,592 lithotomy cases was
about 1 in 9, or 11 per cent.   There were, however,
51 cases discharged as "otherwise" than "cured,"
"relieved," or "died"; and it may be presumed that
these, or the great majority of them, were taken away
by their friends in a moribund condition to die at
home.   If these be added to those dying in hospital,
the total mortality will be about 13 per cent., or 1 in 8
nearly.

I regret I am unable to give the statistics of mortality
according to age.   As the mortality from lithotomy
increases in proportion with the age of the patients
operated on, these would have the effect of bringing
out more clearly the great diminution in mortality from
litholapaxy as compared with lithotomy.   The only
Administration Report which gives the statistics of
mortality according to age is that for the N.-W.
Provinces and Oudh.   Taking the latest statistics for
these Provinces (as being the most likely to be correct),
those for 1883, we find that, amongst 987 cases of
lithotomy performed in that year, the mortality up to
the age of 20 was 5·1 per cent., or nearly 1 in 20 ;
between the ages of 20 and 40 years, 10·7 per cent., or
about 1 in 9·5 ; and above 40 years, 31·9 per cent., or
nearly 1 in 3.

The impression above referred to is, therefore, in-
correct.   The mortality of 1 in 8 nearly, or about
13 per cent., occurring amongst 2,592 lithotomy
operations performed in Indian hospitals in one
year, on patients of all ages, is, practically, the same

as that recorded by Erichsen as occurring amongst Europeans.

In my own practice I have performed 321 operations for stone in the bladder. Amongst these were 171 adult males and 1 adult female, 143 male and 6 female children. Litholapaxy was performed on 125 adults (124 males and 1 female), with 5 deaths ; and on 3 female children, with no deaths. In 3 female children the calculi were successfully removed by rapid dilatation of the urethra ; this was before the introduction of litholapaxy. There were 47 adult males treated by lithotomy, with 9 deaths. In the 143 cases of stone in male children, lateral lithotomy was the operation performed on all ; and *amongst these there was no death.*

There were, therefore, 125 adults treated by litholapaxy, with 5 deaths, or 4 per cent., against 47 adults treated by lithotomy, with 9 deaths, or about 19 per cent. This is not, however, a fair method of instituting a comparison, as, since the introduction of litholapaxy in my practice, only cases unsuitable for this operation were treated by lithotomy. Previous to my practising litholapaxy I had 33 cases of lithotomy in the adult, with 6 deaths, or a mortality of over 18 per cent. Since then I have operated on 139 adults, with 8 deaths (125 by litholapaxy, with 5 deaths, and 14 by lithotomy, with 3 deaths), or $5\frac{3}{4}$ per cent. This is the proper mode of comparing the results of the two operations ; and it will be seen that the introduction of litholapaxy into my practice has been attended by a vast diminution in mortality.

Such, then, are the statistics of the operation, so far as my experience of it goes. The remainder of this work will be devoted mainly to a consideration of the chief difficulties met with in the operation. But, before entering on this subject, I wish to refer to a new method of diagnosis of stone recently brought to the notice of the profession by me.

A NEW METHOD OF DIAGNOSIS OF STONE.

In the great majority of cases that come under observation, sounding for stone is a simple proceeding. Almost any sound will detect a large stone ; and, unfortunately, the greater proportion of stones met with in India are large when patients present themselves for treatment. The form of sound I almost invariably use at first is a No. 8 or 10 of the shape illustrated in Fig. 14. A sound of this kind is very easy of introduction, causing the patient little or no pain ; and, in addition to detecting the stone as a rule, affords collateral information regarding the capacity of the urethra.

Should I fail in detecting a stone by this, I have recourse to Sir Henry Thompson's sound (Fig. 18). The short curved beak of this latter enables one to rotate the instrument in the bladder, and institute a search in all directions, especially behind the prostate, where a stone often lies concealed and is passed over by the ordinary sounds. Being hollow, by the removal of the plug at the end of the handle, water can be allowed to flow from the bladder without withdrawing the sound, and the viscus thus searched with varying quantities of fluid in its interior. An approximate

estimate of the size of the stone may also be ascertained
by means of the small clip, or collar, which
slides along the shaft. The method of using
it is this : " Introduce the sound, feeling the
stone as the end passes over it by a succession of delicate taps, until you have placed
the end of the instrument directly beyond
the farther or distant extremity of the calculus as it lies in the bladder ; this done, slide
the collar down the shaft to the end of the
penis, so that it touches the external meatus.
Now draw the end of the sound outwards
over the stone, delicately tapping as before,
until you have reached its near extremity,
which is most likely close to the neck of the
bladder. The distance of the collar from the
end of the penis is the diameter of the stone
in the direction passed over" (Thompson).
Of this sound I cannot speak in terms of too
high praise. Frequently, when I have failed
with all other varieties of sound in detecting
a stone, I have brought its presence to light
by means of Sir Henry Thompson's.

I find, however, that even with Sir Henry
Thompson's sound a small calculus, lying in
some peculiar position in the bladder, may
evade detection ; and experience teaches us

Fig. 18. that a patient is frequently sent away from
hospital with a stone in his bladder when an opinion
to the contrary has been expressed. The detection of
such small calculi, before they grow into large ones,

is of vital importance, as their removal by the modern method is a very simple proceeding, and unattended by danger.

In the *Indian Medical Gazette* of March 1884 I called the attention of the profession to a new method of diagnosis for small calculi, by means of the aspirator and canula. The method of employing it will be best indicated by a case from actual practice.

CASE LIV.—On August 1, 1883, a Hindoo male, aged 50, came to hospital with symptoms of stone, the most marked of which were sudden stoppage of the flow of urine and increased frequency of micturition. After a most careful exploration of the bladder by sounds of various kinds, including Sir Henry Thompson's, no calculus could be detected. I felt certain, however, from the symptoms, that there was a small stone present, and determined to employ the aspirator for the purpose of diagnosis. I introduced a No. 14 catheter, and applied the aspirator. After going through the performance of pumping water into the bladder and exhausting it once or twice, a distinct click was heard. The canula was withdrawn, the lithotrite introduced, and the stone crushed. The fragments weighed 11 grains only. Next day the man was walking about quite well.

In the above case a most careful search was made by sounds of various kinds, but no calculus could be detected till the aspirator was employed, when a distinct click was heard during the exhaustion of the water, due to the calculus being carried with force against the eye of the canula by the outward stream. The sound of the fragments clicking against the canula during aspiration in the operation of litholapaxy first

F

suggested to me this mode of diagnosis, and I now always employ it when the symptoms of stone are present, and the sound fails to detect one. In this way I have detected several small calculi.

The practical advantages of this simple mode of diagnosis for small calculi are borne testimony to by several of my fellow-labourers in this department of surgery, and especially ' by Mr. Reginald Harrison, of Liverpool, in a work* recently published by him. Mr. Harrison remarks that " where the bladder has lost its shape, either by the encroachment of the prostate or by the development of saccules, the detection of a small calculus is often attended with considerable difficulty, and may be doubtful." In cases of this nature, and others " where stone is suspected, but cannot be readily detected on the introduction of a sound," Mr. Harrison says that he has, since reading my paper in the *Indian Medical Gazette*, systematically employed my method of diagnosis ; and adds : " I have by this instrument (the aspirator-canula) been enabled in at least a dozen instances, not only to detect the stone without distressing the patient, but at once to remove it."

A typical example is given by Mr. Harrison :—

" In a recent case of irritable bladder with cystitis. which I saw in consultation with Mr. Richard Williams, where we had reason to suspect stone in the bladder, the process was adopted, and may well serve as an illustration. We first carefully examined the bladder, under ether, with a sound, but

* " Further Observations on the Treatment of Stone in the Bladder," by R. Harrison, F.R.C.S., 1885, p. 2.

failed to detect a stone in consequence of the great irregularity in the shape of the inferior portion of the viscus. The aspirator-catheter was substituted for the sound, when calculi were at once found clicking against the eye of the instrument. In this way, not only was the presence of stone demonstrated, but these were readily removed, when we were able to declare that the viscus was free."

And he adds : " By this simple process the operation of sounding has been rendered more certain, and freer from those consequences which are sometimes inseparable from the more usual method when required in the case of abnormally shaped bladder."

Not alone may the aspirator be usefully employed for diagnostic purposes, but by means of it a small calculus or number of small calculi may be removed entire, without the necessity of having recourse to the lithotrite at all. Case LXXVI. is a practical illustration of this.

CASE LXXVI.—The patient, a warder in the Moradabad gaol, aged 50, had been passing gravel for two or three years, when suddenly one day his urine ceased to flow. He went to the hospital assistant, who passed a catheter and relieved the retention. Next day he had retention again, when he consulted me at my morning visit to the gaol. I sent him to the Civil Hospital, placed him under the influence of chloroform, and passed a No. 18 catheter at once with the greatest ease. The aspirator was then applied, and the click of a stone heard. The calculus, which weighed only 15 grains, passed into the apparatus, and was removed in this way without the use of the lithotrite. Next day the warder returned to his work quite well, having suffered no unpleasant symptoms.

F 2

I had for some time previously contemplated the removal of a small calculus in this way by the aspirator alone : but this was the first opportunity I had of putting my idea to a practical test.   Since then I have removed calculi in four or five instances in this manner without the aid of the lithotrite.   For instance, in Case LXXXI. I removed twenty-seven small hard uric-acid calculi, the whole weighing half a drachm, all passing through a No. 16 canula into the receiver ; and in Cases XCVIII. and CIX. small calculi were both diagnosed and removed by the aspirator alone.

DIFFICULTIES AND COMPLICATIONS MET WITH:

ILLUSTRATIVE CASES.

OF all the complications met with in the treatment of stone by litholapaxy, the most difficult to deal with, as might be expected, is the presence of organic stricture of the urethra. To permit of the large instruments used in the operation passing through the urethral canal, the stricture must first be disposed of, either by internal urethrotomy or dilatation. The following illustrative cases will show the manner in which this complication may be successfully dealt with :—

CASE LI.—A Hindoo male, aged 65, admitted on the 5th of July 1883, with symptoms of stone, the presence of which was confirmed by the sound. The symptoms had existed three years, and the patient was extremely weak. He was carried to hospital in a bed, and was unable to stand, or even to sit up, without aid. There was excruciating pain in passing water. He had to pass urine every half-hour or so, only a few drops coming away at a time. The urine was blood-stained and mixed with pus and shreds of lymph. There was much albumen present. On passing the sound it was ascertained that there were two strictures present, one an inch behind the glans, and the other four inches from the meatus, through which a No. 6

sound only would pass. The patient was suffering from fever, and so extremely weak that I was afraid to undertake any operation ; but he was admitted to hospital and placed under preliminary treatment. On the 8th of July there was very little improvement, but I determined to operate. The patient being anæsthetized, the first stricture was divided by means of a long narrow scalpel passed along a director, and the meatus, which was narrow, cut at the same time. The deep stricture was then cut by means of Sir Henry Thompson's urethrotome (Fig. 19). A full-sized lithotrite was then passed, the stone caught and crushed, and the débris removed through a No. 18 canula, which was passed without difficulty. The stone, which was uric acid, weighed $1\frac{1}{2}$ drachm. The bladder felt sacculated ; and a large quantity of filthy pus and flakes of lymph were brought away by the aspirator with the fragments of stone. The operation lasted only ten minutes. A full-sized gum-elastic catheter was then tied in. July 9th.—Patient very weak ; suffered from high fever last evening ; urine blood-stained and mixed with pus. July 10th.—Fever again last night ; very weak, and wanders in his conversation. July 11th.—No fever ; patient much better—sitting up in bed ; urine clear ; catheter removed. From this time convalescence was rapid, and he was walking about on the 14th of July. Discharged cured on the 20th of July.

Fig. 19.

CASE CV.—A Hindoo male, aged 50, admitted into the Bareilly Civil Hospital, on the 10th of September 1884, with symptoms of stone, which had existed four months. Micturition very painful and difficult, and frequent stoppage of water. A small sound passed, and stricture detected at the membrancous portion of the urethra. The patient being chloroformed, internal urethrotomy was performed by the urethrotome. A medium-sized lithotrite was passed, and the stone crushed. The débris was removed through a No. 14 canula, and weighed 15 grains only. A gum-elastic catheter tied in. This patient made a rapid recovery, and was discharged cured on the 16th of November.

CASE CXIV.—A Hindoo, aged 45, admitted into the Bareilly Civil Hospital, November 31, 1884, suffering from symptoms of stone, which had existed one year. Contracted gonorrhœa twelve years previously. On passing a sound, a small stone was felt, and the urethra was found contracted at the membraneous portion, admitting a No. 10 sound with difficulty. Patient anæsthetized, and a series of sounds (Fig. 14) from No. 10 to No. 16 passed rapidly one after another. A medium-sized lithotrite was then passed easily, and the stone caught and crushed. The calculus was mixed, partly uric acid and partly oxalate of lime; and the fragments weighed 1 drachm and 20 grains. Canula No. 14 was used. Evening.—Patient in great pain; passing urine by means of catheter only, which was three times introduced by my assistant-surgeon, and eventually a gum-elastic catheter tied in. Poultices to the hypogastrium, and hot fomentations. Dover's powder, 10 grains, internally. November 1st.—No pain; passing water freely through the catheter; no fever. Patient rapidly recovered, and was discharged on the 12th of November.

The preliminary treatment of the stricture resolves itself into either of two methods—internal urethrotomy or dilatation. The former is the mode I prefer as a rule. But cases will occur in which the latter may prove the more suitable. When dilatation is had recourse to, this may be effected by passing a series of conical steel sounds (Fig. 14) rapidly one after another,

through the stricture, or by means of Holt's dilator
(Fig. 20). For my own part, I have now practically
abandoned this latter method of dealing with stricture
in favour of internal urethrotomy ; but there are many
surgeons who still cling to this
method, and I therefore mention it. Should the method
of dilatation by means of conical steel sounds rapidly passed
in succession be had recourse
to, it will often be well, especially when the stricture is
narrow, to commence dilatation of the stricture a few days
before the operation, by passing and tying in gum-elastic
catheters of successively larger
sizes till No. 10 or 12 is
reached ; and then, on the day
of operation, completing the
dilatation by means of the
steel sounds rapidly passed.
Mr. G. Buckston Brown has
recorded[*] two very interesting cases in which this procedure was successfully employed.

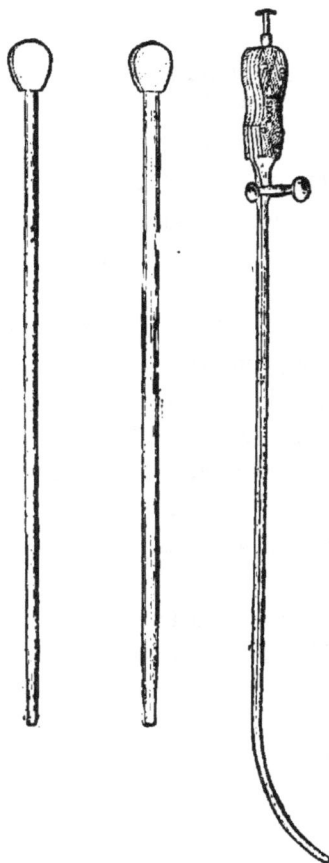

FIG. 20.

In all cases in which a stric-
ture of the urethra has been the subject of treatment
immediately before the operation of litholapaxy, a soft

* *Lancet*, November 10, 1883, p. 810.

gum-elastic catheter should be passed at once after the operation, and tied in for a few days. The omission to do this at first in Case CXIII. was the cause of some trouble.

Enlargement of the prostate is a complication which, contrary to what might be expected, as a rule offers little or no obstruction to the performance of litholapaxy. In passing the instruments over the enlarged prostate, a little extra manipulation may be necessary, and this can only be learnt with practice. When obstruction is met with at the prostatic portion of the urethra, I find the manipulation of depressing the handle of the lithotrite between the thighs, and pushing it on with a slight rotatory motion in the direction of the axis of the body, frequently successful in entering the bladder. Should this fail, it will be necessary for the surgeon to change the right side of the patient for the left, and, by means of the forefinger of the left hand in the rectum, endeavour to guide the point of the instrument over the obstruction into the bladder.

When enlargement of the prostate is accompanied by atony of the bladder—as it frequently is—care must be taken to draw the urine off two or three times daily after the operation by means of a catheter, or a soft gum-elastic catheter may be tied in and the urine allowed to flow by this for a few days. The following case illustrates some of the difficulties met with : as also the after-treatment. Case CXXIII., which will be subsequently given under the head of large calculi, also illustrates some of the difficulties arising from enlarged prostate, as well as the means by which they may be overcome.

Case XXXIV.—A Mahomedan mason, aged 85, was admitted on the 15th of April 1883, with symptoms of stone in the bladder, which had existed two years. There was great pain in passing urine, which only passed in small quantities at a time, frequently repeated. On passing a sound, the presence of a stone was confirmed, and the existence of a greatly enlarged prostate also ascertained. On passing a catheter, a large quantity of "residual" urine was drawn off. The patient was emaciated, extremely feeble, and almost in a dying state. Still, he was at once anæsthetized, and litholapaxy performed. There was considerable difficulty at first experienced in passing the lithotrite over the prostate. This was obviated by passing the instrument on the left side, with the finger in the rectum as a guide. The operation lasted twenty minutes, during which $6\frac{3}{4}$ drachms of a very hard uric-acid calculus were removed. The lithotrite had to be introduced four times, and a No. 16 catheter as often. Evening.— Retention of urine ; catheter passed and water drawn off ; patient suffering from high fever, very weak. 16th.—Fever less ; retention of urine continues ; catheter passed every six hours. 19th.—On passing the catheter it grated against a fragment of stone. Chloroform given, and the fragment crushed and evacuated by the aspirator ; weight, 40 grains. From this time the patient made a rapid recovery, putting on flesh and picking up strength, and was discharged cured of stone on the 26th of April. The enlarged prostate and atony of the bladder still continued of course.

Such, then, are the means by which the complications of stricture of the urethra and enlargement of the prostate may be overcome as a rule. We must not, however, expect to be always successful in such cases. It may happen, even when a stricture has been successfully dilated, or a large conical steel sound passes with ease over an enlarged prostate into the bladder, that a lithotrite cannot be passed with any amount of manipulation. The following case is a typical illustration of this :—

A Rajput, aged 45, admitted into the Mussoorie Hospital on the 27th of April 1885, with the usual symptoms of stone, which had existed 1½ year. On passing a No. 10 steel sound, it was found that there was a constriction at the prostatic portion of the urethra, some force being necessary to pass the sound into the bladder, where a calculus was detected. On the 28th I passed, and tied in, a No. 10 elastic catheter. On the 29th, when this was withdrawn, a No. 12 sound passed easily. A No. 12 elastic catheter was then tied in, and allowed to remain there till the 1st of May, when I tried to perform litholapaxy. The patient being anæsthetized, I endeavoured to pass lithotrites of all sizes over and over again; but, though a No. 14 conical steel sound could be passed with ease, I failed in all my attempts to pass even the smallest sized lithotrite. The beak of the instrument invariably hitched in front of the prostate, and got caught in a kind of pouch on the left or right. As a last resort, I had to perform lithotomy, a dumb-bell shaped stone weighing 6½ drachms being removed. The prostate, when felt through the incision in the perineum, was found to be enormously enlarged and irregular, which was remarkable for a patient of that age, though I have noticed that enlargement of the prostate sets in much earlier in life in natives of India than amongst Europeans. The patient made a good recovery, and was discharged on the 13th of June.

A difficulty is sometimes met with, both in passing the instruments and catching the stone, when the calculus lies stationary, growing partly in the bladder and partly in the prostatic portion of the urethra. From one's experience of lithotomy, the difficulty of managing such cases may be easily imagined. Every lithotomist of any experience must have come across cases in which an irregularly elongated calculus lies with its main portion, or body, in the bladder, and a small elongated head in the prostatic urethra, the two portions being united by a neck corresponding with the vesical orifice of the urethra. Such a calculus must be displaced from its position backwards into the bladder before being crushed. The manner of dealing with such calculi will be best illustrated by a case from actual practice :—

CASE CXXI.—This case I saw in consultation with Surgeon-Major Fasken, at Dehra Doon, with whose kind permission I performed litholapaxy. A Hindoo, aged 32, admitted on the 17th of March 1885, with the usual symptoms of stone, which had existed five years. Patient very thin and weak ; passes urine in drops with great pain. On the 20th, chloroform being given by Dr. Fasken, I operated. On passing a full-sized sound, the stone was met with at the neck of the bladder, and obstructed its advance, but by manipulation the sound was passed into the bladder *over* the stone. The same difficulty was experienced in passing the lithotrite, and the stone could not be grasped. The lithotrite was therefore withdrawn, a No. 18 canula introduced as far as the end of the stone lying in the prostatic urethra, the aspirator applied, and water pumped with force into the bladder. By this manœuvre the stone was displaced backwards into the bladder by the force of the stream, the prostatic urethra grasping the stone being at the same time dilated by the water, and so loosening its hold on the stone. The stone was then grasped by the lithotrite, and soon

disposed of. The operation lasted twenty-three minutes, the débris of the stone, which was mixed uric acid and phosphates, weighing 5½ drachms. There was some pain in the urethra for a day or two, with some dribbling of water, but the patient was discharged on the 3rd of April perfectly well.

As I have already mentioned, there were several calculi weighing 2 ounces and upwards removed by litholapaxy, and three calculi weighing each about 3½ ounces. The removal of large calculi of this kind is no light or easy task, and will be found to call forth all the resources of the surgeon. The chief difficulties met with, and the means by which they may be overcome, are well illustrated in the two following cases, which I will give in detail as they involve many points of interest :—

CASE XVI.—Illahi Buksh, a Mahomedan, aged 60, was admitted into the Moradabad Civil Hospital on December 3, 1882, with all the symptoms of stone in the bladder, which had existed eleven years. On admission, the patient could only pass urine in drops continuously throughout the day and night, and the passage of urine was attended with great pain. His penis and foreskin were hypertrophied from the patient's constantly rubbing the organ to relieve the pain and irritation. A urethral calculus was felt in the fossa navicularis. When the patient tried to pass urine he had to rub and pull the penis, and in this way push the urine past the calculus in the urethra. The urine was mixed with pus and blood. The fæces passed were ribbon-shaped, due to pressure of the stone in the rectum. On passing the finger into the rectum, a large stone could be felt in the bladder. The patient's health was very bad. He was pale, thin, weak, and anæmic, and he had a pinched, anxious expression, the result of long suffering.

On December 6 I performed litholapaxy, the urethral calculus
having first been removed after slitting the floor of the urethra
slightly. The operation lasted sixty-six minutes, and the
débris weighed 3¼ ounces, the calculus being a hard uric-
acid one. Considerable trouble was at first experienced
in grasping the stone owing to the contraction of the walls
of the bladder on it. This was obviated by injecting
water into the bladder. The lithotrite was introduced at
least a dozen times, and after each crushing a large quantity
of débris was washed out through a No. 18 canula. With the
exception of slight pain in micturition during the first day or
two, the patient had no after-trouble. He made a rapid
recovery, and on December 15, when discharged from the
hospital, the following entry in my note-book describes his
condition :—" Patient now rid of all bladder symptoms.
Urine quite clear ; bladder retains a large quantity at a time.
Has grown fat and strong. Says he has not been so well for
several years. This man was a miserable creature on admission
to hospital ten days ago, and now leaves it in excellent health."

CASE CXXIII.—A Mahomedan male, aged 60, admitted into
the Mussoorie Hospital, August 29, 1885, with symptoms of
stone. On passing a sound, the presence of a large calculus
was confirmed. The patient stated that the symptoms had
existed thirty years, but that they were extremely aggravated
during the last three years. He was very weak and
emaciated, and scarcely able to stand up. He agreed to an
operation, provided that it should not be a cutting one.
On the 30th of August I performed litholapaxy. Surgeon-
Major W. Murphy, I.M.S., and Surgeons A. Kavenagh and
C. R. Tyrrell, Medical Staff, being present. Chloroform was
given by Dr. Tyrrell. A large sound, No. 14, was first passed
with ease, after slitting the floor of the meatus slightly. I
then attempted to pass the largest lithotrite, but its beak
was arrested by a pouched condition of the urethra near the

neck of the bladder. The medium-sized and smaller lithotrites were then tried, but with a similar result. Time after
time the large steel sounds were passed easily, but any sized
lithotrite I failed to introduce. Eventually, after no less than
twenty-five minutes had been wasted in the attempts to pass the
lithotrites, the medium-sized lithotrite was passed successfully.
This was accomplished by passing the instrument as far as
the part at which the hitch took place, then depressing
the handle between the thighs, and pushing the instrument
on with a rotatory motion in the direction of the axis of the
body. The medium-sized lithotrite could not close on the
stone, so it was withdrawn, and the large lithotrite then
easily passed. But the stone could not be grasped owing to
the walls of the bladder contracting on it. It was therefore
withdrawn, and 4 or 5 ounces of water introduced by
means of the aspirator and canula. The stone was then
caught by the large lithotrite, but failed to " lock," owing to
the large size of the stone. The long diameter of the stone
was then changed for a shorter, when it was found that the
lithotrite locked with ease. The calculus was extremely hard
and tough, and required all the force I was capable of to crush
both it and the fragments. The lithotrite had to be introduced several times, and the evacuating catheter, No. 18.
as often, before the whole of the stone was removed. The
fragments, when moist, weighed nearly 4 ounces, and when
dry, 3 ounces 2½ drachms. The operation lasted one hour in
all, of which twenty-five minutes were wasted in unsuccessful
attempts to introduce the lithotrites at first. Once the stone
was grasped, the large lithotrite soon disposed of it. There
was considerable bleeding from the urethra during the early
manipulations, though great care was taken to use no force.
Immediately after the operation the patient was extremely
weak and almost pulseless. On consciousness returning.
stimulants were given, and the usual after-treatment had

recourse to. The stone was partly uric acid and partly oxalate of lime. August 31st.—Patient passed a good night ; says that he has not had such a good night's rest for thirty years! urine passing freely, with some pain, and blood-stained. On the 1st of September there was some pain and tenderness in the region of the bladder, which, however, yielded to hot fomentations. The patient's temperature never went above 100° F. He suffered for a few days from a discharge of pus, which came from the prostatic portion of the urethra. This was evident, as, on washing out the bladder with astringent lotions, no pus came from the viscus. Probably the urethra at this point was slightly injured during the attempts to pass the lithotrites. The patient put on flesh, and was discharged cured on the 24th of September, though for several days previously he was walking about the hospital inclosure.

I have considered it advisable to record both these cases in detail, as they present several features of interest. These are, as far as I am aware, the largest calculi that have ever been successfully removed from the bladder by the one-sitting operation. The largest recorded by Sir Henry Thompson* is 2¾ ounces. The amount of manual labour required in crushing large calculi of this nature is something excessive. On each occasion I felt completely exhausted after the operation. My hands were blistered from the lithotrite, and my arms ached for days subsequently. Great patience will be required in the various manipulations before the stone is caught. As a rule, when the stone is large, the walls of the bladder hug it closely, so that

* "Diseases of the Urinary Organs," sixth edition, p. 72.

the manœuvre referred to in these two cases, of injecting water to separate the walls of the bladder
from the stone, must be had recourse to before the
stone can be grasped by the lithotrite. It will sometimes be found also, in dealing with large calculi, that,
though the lithotrite will not "lock" should the stone
be first grasped by the long axis, it will do so if this
is changed for the short axis of the stone. This
manœuvre should always be tried before abandoning
the case as unsuitable for litholapaxy.

I cannot too strongly advocate the desirability of
removing the whole of the calculus at one sitting. This
is the essential principle of the operation. This fact
seems to be lost sight of by some surgeons, for I have
noticed that in some of their recorded cases a second
and even third operation was necessary. An operation
prolonged over several sittings in this way involves all
the dangers of the old operation of lithotrity, and
ceases to be litholapaxy in my acceptation of the term.

The following case illustrates a difficulty, though a
slight one, which I have only once met with, but which
I consider of sufficient interest for record here :—

Case CXXV.—A Brahmin, aged 45, was admitted into
the Mussoorie Hospital on the 6th of September 1885, suffering from the usual symptoms of stone, which had lasted 2½
years. On passing a sound, two calculi were diagnosed. On
the 7th I performed litholapaxy, chloroform being given by
Surgeon Tyrrell, and Drs. Whittall, Fiddes, Murphy, and
Burlton being also present. After slitting the floor of the
meatus, the large lithotrite passed with ease, and the calculi,
which were soft, were easily crushed. The No. 13 canula

G

passed with facility. On applying the aspirator and attempting to inject water, however, the apparatus failed to work. I made several attempts to pump in water, but to no effect. I then withdrew the canula, and found that it was full with soft, mortar-like débris, which, when removed by tapping the canula against a vessel, remained in the form of a cast of the canula, like a piece of thick macaroni. After this the operation proceeded without further difficulty, and the patient made a speedy recovery. In fact, he was sitting in the hospital inclosure next day. The débris consisted mainly of phosphates, and weighed $4\frac{1}{2}$ drachms when dry. When the obstruction to the ingress of water occurred, I was at a loss to account for it, and first imagined there was spasm of the bladder, such as is sometimes met with even when profound anæsthesia exists; but on withdrawal of the canula the cause was apparent. The débris being in a plastic, semi-fluid state, rushed into the canula immediately it was introduced, and formed a cast of it.

Such, then, are the main difficulties and complications met with in the operation. I will now give a series of interesting cases, many involving difficulties, with practical remarks thereon.

INTERESTING CASES,

WITH PRACTICAL REMARKS.

FROM the preceding pages it will have been observed that several patients of 80 years and over had large calculi successfully removed. The oldest patient on whom I have operated was a Mahomedan aged 96, from whom I removed successfully a hard uric-acid calculus, the débris of which weighed 9½ drachms, the operation lasting one hour. An interesting feature in this case was, that till about a month before coming under observation he had enjoyed excellent health and was untroubled by any urinary symptoms. The details of this case are interesting, so I give them.

CASE LXIX.—This was a case in private practice. The patient, a Mahomedan of Moradabad, stated that he was close on 100 years of age, and by calculation he appeared to be 96. He was a dried-up, withered creature, without a tooth in his head, and consisting almost of skin and bone only. Several of his sons were living, and one of them looked 70 years of age. Till about a month before coming under my treatment he had enjoyed good health, and used to walk about the bazaar daily. He was suffering from well-marked symptoms of stone, especially great pain in passing urine. He was very weak, and unable to leave his bed.

On October 28, 1883, I performed litholapaxy, the débris weighing 9½ drachms, and the operation lasting one hour. The patient made a rapid recovery, and was able to walk about on November 13. Five months afterwards I had the pleasure of showing this old gentleman to Surgeon-General W. Walker. He was then in excellent health, so much so that Dr. Walker, writing of him at the time, amusingly says : " He must certainly be 90, and looks as if he might live thirty years more, and then do service as an old rail."

My experience is that old patients bear the operation much better than young men, and that in them it is less likely to be attended by urethral fever. The explanation of this is to be found, I have no doubt, in the fact that the mucous membrane of the bladder and urethra gets less sensitive as life advances, and also that the urethra is more capacious in old than in young men.

The rapidity with which a stone may be removed from the bladder by the modern operation will vary according to circumstances. The crushing power of the lithotrite, the efficiency of the aspirator, the calibre of the urethra, the shape and capacity of the bladder, the size and hardness of the stone, the dexterity and experience of the operator, are all factors that will have to be taken into consideration. The longest time I have spent over the operation was 1½ hour, and this was in the case of an old man with an irregular, rigid, contracted bladder, the calculus weighing 2 ounces 1 drachm. The details of the case are as follow :—

CASE LXX.—A Mahomedan, aged 80, admitted into the

Moradabad Hospital on the 21st of November 1883, with symptoms of stone which had existed some four years. He was very thin and weak, and micturition was extremely painful. I had sounded this man a week before his admission to hospital, and found a large stone. The patient being anæsthetized, I performed litholapaxy. The calculus was a hard uric-acid one, and required several introductions of the lithotrite and evacuating catheter (No. 16) before it was completely removed. Owing to the bladder being contracted and its inner surface very irregular and rigid, great difficulty was experienced in catching the last fragments by the lithotrite. For this purpose a shallow, flat-bladed instrument was employed. The aspirator and canula were used for bringing the fragments close to the neck, on the trigone. This was effected by introducing the canula, pumping in water and then exhausting it, when a fragment, too large to pass, was carried against the eye and held there by the suction force of the exhaust. The apparatus was then withdrawn till the end of the canula reached the neck of the bladder, and the fragment detached by the least pressure of the hand on the bulb, thus relieving the suction force of the apparatus. The fragment being thus placed in the most favourable position, the lithotrite was introduced and the fragment secured and crushed. The operation lasted one hour and a half, and the débris weighed 2 ounces 55 grains. The old man made a rapid recovery, and was discharged on the 4th of December.

Given, on the other hand, a patient with a capacious urethra and healthy bladder, it is wonderful the rapidity with which a large stone may sometimes be removed by the modern instruments. The following case is a good illustration of this :—

CASE XCI.—A Hindoo, aged 35, admitted into the

Bareilly Civil Hospital on the 21st of June 1884, with the usual symptoms of stone, which had existed 2½ years, the pain being excessive during the last few days before admission. The patient being anæsthetized, litholapaxy was performed. It was necessary to incise the floor of the urethra slightly, and then my largest lithotrite passed with ease. The calculus was phosphatic, not very hard, and the fragments weighed exactly 2 ounces. The operation lasted only seventeen minutes, only two introductions of the instruments being required. This was the first occasion on which I used a large new lithotrite made for me by Weiss, and I was much impressed by its wonderful crushing power. Canula No. 18 was used. The patient recovered without a bad symptom, and was discharged. on the 30th of June.

Not less interesting than the facility with which large calculi may sometimes be removed by the modern operation is the rapidity of the cure that ensues in some cases, even when the stone is large and the operation prolonged over a considerable period. I will give details of two such cases.

CASE LXVIII.—A Hindoo, aged 55, came to the Moradabad Hospital, on the 20th of October 1883, suffering from the usual symptoms of stone, which he stated had only troubled him for six months, during which time he had great difficulty in passing water. On the 26th of October I performed litholapaxy, the Sanitary Commissioner, Deputy-Surgeon-General C. Planck, being present. The calculus was very tough and hard to crush, and the operation lasted thirty-four minutes. The stone was partly oxalate of lime and partly uric acid, the débris weighing 1 ounce 35 grains. Catheter No. 18 passed easily without incising the meatus. Next day the patient was walking about the hospital as if no

operation had been performed. On the 30th of October, the
day on which the man was discharged, he presented himself
at Dr. Plauck's camp. Dr. Plauck was very much struck with
the rapidity of the cure, having seen the man operated on four
days before only. The patient was in high spirits, and gave
us an amusing account of the various expedients he used to
have recourse to in order to pass urine before the operation.
He said that he used to lie on his back, then on his belly,
sometimes on one side and then on the other, but frequently
without avail. When these positions failed to give him relief,
he had recourse to standing on his head with his legs in the
air; but even then he frequently could not pass water!

CASE LXXX.—A Mahomedan, aged 45, admitted 11th
March 1884, with stone. The symptoms had existed three
years. Litholapaxy performed by me, Surgeon S. Thomson,
B.M.S., being present. The operation lasted half an hour,
during which the instruments were introduced four times.
The stone was a hard uric-acid one, and the débris weighed
1 ounce 2½ drachms. The patient was walking about the
hospital next day, quite well, and desirous of going home.
He was discharged on the 17th of March, having had no bad
symptoms.

In practice in India one is frequently amused by the
naivete of the peasantry and the rude devices they
frequently have recourse to, intuitively, for the relief
of disease. These remarks apply partly to Case LXVIII.,
above recorded. Of a somewhat similar nature was
Case XXIX., in which, with the assistance of Dr. Wilson,
medical missionary of Ghurwal, I removed on the 5th of
March 1883, a calculus weighing 2½ drachms from a
hillman, aged 46 years. This man informed us that
during the previous year he had been in the habit of

employing a thin and pliant bamboo twig, which he produced, to assist him in passing water. This he passed through the urethra, and by means of it pushed back the stone from the neck of the bladder preliminary to passing urine. He then went through a process resembling, in many respects, the milking of a cow. He pulled his penis with force, putting it on the stretch, then relaxed it, when some urine passed away with a rush. This process was repeated again and again till his bladder was emptied.

I have already stated that no selection of cases was made, and that patients in the very worst conditions of health were frequently operated on. This will have been apparent from some of the cases already given in detail. In order to illustrate what the modern operation is capable of in rescuing from death many sufferers on whom no cutting operation could be entertained, I am tempted to give details of the following cases :—

CASE LVI.—A Mahomedan, aged 85, was brought to the Moradabad Hospital on the 8th of August 1883, with symptoms of stone, which had existed five years. The patient was extremely exhausted from his sufferings; he was a mere skeleton, unable to walk, and had to be carried in a bed. The pain in the bladder was very severe, and micturition frequent, only a few drops of urine coming away at a time. There was no albumen. The patient was so weak that no operation could be entertained at once, so he was kept in hospital for preparatory treatment. On the 12th of August there was scarcely any improvement; but, as the man was clamouring for the operation, chloroform was given and litholapaxy performed. The operation lasted nearly an hour, and the débris of the calculus, which was a hard uric-acid one, weighed nearly 1¾ ounce. The urethra admitted the largest lithotrite and canula, No. 18. With the exception of slight fever for the first day or two, and some pain in micturition, there was not a bad symptom. On the 18th he was walking about the hospital, and expressed himself as " feeling forty years younger than before the operation." This man was discharged on the 23rd, in good health. Some months after, when out in the Terai on a shooting trip, I met him in his native village, and he was then quite well.

CASE CVII.—Golam Abbass, a respectable Mahomedan of Bareilly city, aged 45, came to the hospital, suffering from all the symptoms of stone, which had existed two years. The presence of a stone was confirmed by the sound. Previously to coming under observation, he had undergone " four courses of purgatives " at the hands of a native *hakim*, under the impression that he was suffering from some other disease than stone. The pinched and anxious expression of his face indicated that he had undergone extreme suffering. He was pale, thin, and anæmic ; and his body and limbs consisted almost solely of skin and bone. In fact, he was a living skeleton. The man was unable to walk from pain and debility, and had to be carried in a bed. The assistant-surgeon and subordinates of the hospital considered an operation unadvisable, as the patient was in a dying state. My friend Dr. Corbett, B.M.S., also saw the case, and agreed with me in thinking that, were lithotomy the only alternative, it would be advisable to allow the man to die unoperated on, as he would be likely to die on the operating table from shock and loss of blood. With the kind assistance of Dr. Corbett, I performed litholapaxy on the 24th of September, removing 9 drachms of hard uric-acid calculus, the operation lasting thirty-five minutes. With the exception of slight fever on the day of the operation, the patient had not a bad symptom, and was discharged on the 4th of October, ten days after the operation, in good health. Subsequently, he frequently presented himself at the hospital to make his *salaams*, untroubled by any urinary symptom.

These cases are merely illustrative of many of a similar kind in my practice. In neither instance could lithotomy be entertained, as the patient had not an ounce of blood to spare. It is in cases of this kind that the operation of litholapaxy stands forth in brilliant contrast to that of lithotomy, and creates a marked impression on the mind of the native by the rapidity of the cure and the undisturbed condition of the parts.

I have already referred to the impression which prevails that the mortality from lithotomy in natives is less than in the case of Europeans. There is also an impression prevalent to the effect that natives of India have no fear of the surgical knife. In fact, from the way some people talk and write, it might be almost inferred that a native submits to a surgical operation as a kind of harmless diversion. This impression is altogether erroneous. A native of India will not, as a rule, submit to a surgical operation till all other modes of treatment fail, and he is driven to it through extreme pain, inconvenience, or danger to life. And it is for this reason that such large calculi are met with in this country, and that patients suffering from cancer and other diseases present themselves in hospital at a stage when surgical interference is useless.

Illustrative of the extraordinary ideas natives sometimes get into their heads about surgical operations,

I may mention that some of my patients at first objected to the new operation on the ground that, though they should get rid of the calculus disorder, permanent impotence would result! I have had the curiosity to make inquiries on this point from several of the patients subjected to litholapaxy, seen at varying intervals after the operation, and I need scarcely add that there were no grounds whatever for the ridiculous supposition referred to.

I have already referred to the varying sensitiveness of the urethra and bladder in different individuals, the passage of instruments in some instances being attended by considerable hæmorrhage, and in some cases little or no blood being lost. The latter class are well illustrated in the following two cases, which present some additional features of interest, particularly that of changing the long for the short axis of the stone, when the lithotrite won't lock on it in the former position.

CASE XLII.—A Mahomedan, aged 45, was admitted into the Moradabad Hospital on the 18th of May 1883, with symptoms of stone, which had existed eight or ten years. Health, fair; slight trace of albumen in the urine. The patient being anæsthetized, I performed litholapaxy. The calculus was first grasped by the long axis, but, the lithotrite not locking on it, this was changed for the short axis, when the lithotrite locked. The stone was extremely hard and tough, composed of carbonate of lime, with an oxalate of lime nucleus. The operation lasted one hour and ten minutes, several introductions of the instruments being required. The débris of the stone weighed $2\frac{1}{2}$ ounces. The stone was so difficult to crush that I feared it would be necessary to postpone the completion of the operation to a second sitting, but, by perseverance, I managed to get away all the débris. A great drawback was the fact that only a No. 16 canula would pass. During the whole of the operation there was not a trace of blood in the washings, nor was there any bleeding from the urethral mucous membrane. The patient recovered without a bad symptom, and was discharged in perfect health on the 29th of May.

CASE CXI.—A Hindoo, aged 55, admitted into the Bareilly Civil Hospital, October 25, 1884, with symptoms of stone, which had existed $2\frac{1}{2}$ years. General health, good. On the 26th I performed litholapaxy, Surgeons-Major Knox and Barry and Surgeon Corbett being present. The floor of the meatus was slightly incised to allow the largest lithotrite to pass. On passing the lithotrite, the stone was at once grasped by the long diameter—$2\frac{3}{4}$ inches—but would not lock. The long axis of the stone was then changed for the short, when the instrument locked. I attempted to crush the stone, but three times failed to do so. I then screwed the blades together with all the force I was capable of, rested a second or two, when, suddenly, the stone gave way with a loud report that was audible to all the persons in the room. The fragments were then disposed of with comparative ease. The calculus was composed of uric acid, with an oxalate of lime nucleus. The débris weighed 2 ounces and 45 grains when dried. Catheter No. 18 passed with ease. The operation lasted only twenty-six minutes, there being only three introductions of the instruments. With the exception of the trifling bleeding from the incision in the urethra, there was no hæmorrhage—not a trace in the washings. On the 27th the patient was walking about the hospital, passing urine quite clear, and without pain. Discharged cured on the 5th of November.

The loud report accompanying the crushing of the stone in Case CXI. would suggest the possibility of harm accruing to the bladder by the splintering of the fragments in the case of large stones such as this. In this case, however, the recovery was rapid; and that no injury was done was evident from the fact that there was no bleeding from the bladder at the time, and that subsequently no bad symptoms supervened. I have frequently verified this in other cases; but it is well, in dealing with large and hard stones, to have a considerable quantity of water in the bladder, which acts as a kind of buffer between the fragments and the walls of the bladder. Owing to the calculus in the bladder being saturated with moisture, the fragments do not fly with force, as in the case of a dry stone.

Amongst my cases of litholapaxy a remarkable instance of spontaneous fracture of stone in the bladder occurred, and, as such cases are extremely rare, perhaps its record may not be out of place here. Such a case occurring in the practice of one of those quacks who profess to be able to dissolve stone by the internal administration of secret remedies would probably be the means of making his fortune were the urethral orifice sufficiently large to permit the fragments to pass out.

Case XXX.—A Mahomedan, aged 35, admitted March 19, 1883, with the usual symptoms of stone, which had existed 1½ year. Thirteen days previous to admission all the symptoms became suddenly aggravated, since which time urine had only been passed in drops with difficulty. On placing the patient on the table with a view to passing the sound, I noticed an elongated hard thickening along the course of the urethra, and my first impression was that there was a severe stricture present. Dr. Moran, 6th Bengal Infantry, who kindly assisted me at the operation, remarked that the feeling was like that of urethral calculus, and, on passing the sound, his idea proved to be correct. The floor of the meatus was incised, and no less than 1 drachm of calculus débris removed by the urethral forceps and scoop. The whole length of the urethra was filled with pulverized calculus. A sound was then passed, and fragments detected in the bladder. The lithotrite was introduced, and litholapaxy performed. The débris removed from the bladder, exclusive of 60 grains from the urethra, weighed 125 grains. On inspection of the fragments, it was evident that they belonged to one phosphatic calculus. The patient was discharged cured on the 27th of March.

Amongst the 128 cases in which litholapaxy was performed by me, there were, as already stated, five deaths. The details of these cases are not less interesting and instructive than many of the cases of recovery already recorded, so I give them.

CASE VII.—A Hindoo male, aged 65, was admitted into the Moradabad Civil Hospital on September 21, 1882, with symptoms of stone in the bladder, which had existed three years. There was painful and difficult micturition, with frequent desire to make water ; passing of blood occasionally ; and the urine for some months had been mixed with pus, giving a very offensive smell on standing. There was great enlargement of the prostate. A catheter was passed, and about 10 ounces of fetid urine drawn off. On passing a sound, several small calculi were detected. The man's health was very bad. On September 22 I performed litholapaxy. The operation lasted thirty-five minutes, the lithotrite being introduced four times. The débris weighed 11 drachms. No. 18 canula passed easily—after previously slitting the floor of the meatus slightly—as far as the prostatic portion of the urethra, where some manipulation was necessary to pass it into the bladder. It was evident from the appearance of the débris that there were several calculi, varying in size from that of a pen upwards. Before withdrawing the canula finally, the bladder was washed out with a weak carbolic solution. In addition to the usual after-treatment, a catheter was ordered to be passed morning and evening, owing to the atony of the bladder and enlargement of the prostate that existed. During the first two days little urine passed except through the catheter. The urine continued fetid and sanious ; the bladder was washed out daily, a faint trace of carbolic acid being added to the water. There was no pain, but the patient continued very anxious and depressed, and died on the 27th of September from exhaustion. No post-mortem examination was permitted.

Case XXIV.—A Mahomedan male, aged 48, was admitted on January 9, 1883, with retention of urine, which had existed for thirty-six hours. He had suffered from symptoms of stone for nine months, being much exhausted, with pinched, anxious expression. On passing a full-sized catheter, a small calculus was detected at the neck of the bladder, blocking up the urethral passage. This was pushed back with some force, and the retention of urine relieved. The patient was at once anæsthetized, and litholapaxy performed. The operation lasted only eight minutes ; the débris weighed 20 grains. 10th.—Passed very little urine since yesterday. Pain in region of the bladder. Catheter passed, and a small quantity of urine drawn off. Hot fomentations and hot poultices to the hypogastrium were ordered ; also 1 grain of opium internally every three hours. 11th.—Well-marked peritonitis present. Temperature, 102° F. ; respiration, 44 per minute ; urine, scanty. The catheter was used twice daily. 13th.—No pain. Great distension of abdomen ; patient very weak ; passing a little urine. He died quietly in the evening.

Post-mortem examination.—On opening the abdominal cavity it was found distended with clear serous fluid. The bladder was embedded in a mass of amber-coloured, gelatinous lymph, which broke like cold jelly on application of the fingers ; congestion of the cellular tissue at the base and neck of the bladder ; mucous membrane of the urethra near the neck of the bladder highly inflamed; no congestion of the bladder or kidneys.

CASE XLIV.—A Hindoo, aged 45, was admitted on May 23, 1883, with symptoms of stone, the most prominent of which were agonizing pain in the region of the bladder and passing of blood and pus in the urine. The symptoms had existed eight years, and the patient was in a very weak state. A large calculus was detected. by the sound. On the 25th I performed litholapaxy. Considerable difficulty was experienced owing to the large size and hardness of the calculus, and the operation progressed very slowly, 14 drachms being removed in seventy-five minutes. No more fragments could be felt by the lithotrite, but, on passing the sound, a large fragment was detected high up at the fundus of the bladder. As the patient was very much exhausted, I was afraid to keep him longer under chloroform, so I had, unwillingly, to postpone the completion of the operation to another day. 24th.—He had great pain in the region of the bladder the whole of yesterday, which was controlled by 1 grain of opium administered every three hours. Urine scanty and blood-stained. Temperature, 104° F. ; great thirst. Fever mixture ordered and opium continued. 27th.—Had a severe rigor last night ; fever continued all night ; patient anxious and depressed. Ten grains of quinine were ordered, and to be repeated in the evening. No pain in the abdomen, but great pain in the left hip. 28th.—Abscess forming in the left hip. All the symptoms of blood-poisoning present. Carbonate of ammonia and bark mixture ordered. The patient died at 10 P.M.

Post-mortem examination.—The walls of the bladder were greatly thickened, sacculated, and contracted; in the bladder were found a large fragment of calculus (about 1 ounce in weight) in a sacculus at the fundus, several small patches of inflammation of the mucous membrane in the vicinity of the fragment of stone, and some milky urine. The tissues of the left gluteal region and round the left hip-joint were swollen and infiltrated with dark-red fluid. Cause of death—pyæmia.

CASE XCVII.—A Mahomedan male, aged 26, was admitted into the Bareilly Civil Hospital on August 12, 1884, suffering from the usual symptoms of stone, which had existed six years. During the last three years the symptoms became much aggravated. On admission, micturition was extremely painful and difficult, and the urine was mixed with pus and blood. The patient was in extremely bad health; very thin and anæmic; scarcely able to stand. On August 14 I performed litholapaxy. Chloroform was given by Dr. Corbett, Bengal Medical Service, by means of Junker's inhaler. The calculus was a large hard one, weighing 3 ounces and 1½ drachm, and consisted of oxalate of lime. The operation lasted fifty-two minutes. After slitting the floor of the meatus slightly, a large lithotrite was passed, but, owing to the contraction of the walls of the bladder on the stone, it could not be grasped. The lithotrite was withdrawn, and 4 or 5 ounces of water injected by the aspirator, and then the calculus was easily grasped after a little manipulation. I had to use all the force I was capable of before the stone gave way. The fragments were then disposed of one by one, great force being necessary in dealing with even the smallest fragment. The lithotrite had to be introduced six or seven times, and the evacuator as often, before the whole of the calculus was disposed of. My new large lithotrite did great execution amongst the fragments, the glass receiver of the aspirator

H

being nearly filled each time after withdrawal of the lithotrite. Two or three times during the operation the patient was very faint and almost ceased breathing. and after the operation the pulse was extremely weak. Artificial breathing had to be maintained. occasionally. for an hour after the operation. Evening.—Passed urine freely during the day: the earlier portions were blood-stained. the later clear : no sand or fragments passed : no pain in the bladder. but tenderness along the course of the urethra. Pulse. 90 (very weak). 15th.—Patient very weak : passed three loose stools during the night ; urine passing freely. and quite clear : vomited twice this morning : pulse almost imperceptible : hands and feet cold : drowsy and restless : wanders in his conversation at times. During the course of the day the patient revived a little. but he grew weaker towards evening. and died at 9 P.M.

Post-mortem examination.—Bladder empty (not a trace of sand was present). contracted. and its walls much thickened ; mucous coat was corrugated and slightly sacculated : urethra congested along its whole length : ureters dilated along their whole length. so that the tip of the little finger could be passed into them from the bladder : kidneys extensively diseased : the calyces were much dilated. and the medullary portions had almost disappeared : the cortical portions were hard and pale : pericardium distended with clear straw-coloured fluid.

CASE CXX.—A hillman. aged 61. admitted into the Mussoorie Hospital. January 25. 1855. with symptoms of stone. which had existed two years. The patient was extremely weak and a mere skeleton. He was also suffering from a bad attack of bronchitis. He was placed under preparatory treatment. As the patient threatened to leave hospital unless something was done at once to relieve his sufferings. I had, unwillingly. to perform litholapaxy on the 27th. without waiting for improvement in his general health. Dr. W. R. Henderson present. On passing the lithotrite. the calculus, which

was a small one, could not be found. This was withdrawn and
the canula and aspirator applied, when the click of the stone
was heard against the eye of the former. The lithotrite was
then introduced, and the calculus caught and crushed. The
aspirator had withdrawn it from a sacculus, or remote part of the
viscus, and placed it in a favourable position for the lithotrite.
The débris weighed only 15 grains. It was an oxalate of lime
stone, with sharp spines projecting from its surface. The
patient was placed in a warm room, and rallied well from the
operation, which was a very trifling one. Next morning he was
found in a state of collapse, cold and pulseless. During the
night there was a heavy fall of snow, and the patient was
neglected by his relatives and the hospital servant who attended
on him. The fire in the room was allowed to go out, and no
nourishment was given him during the night. Though he
rallied a little during the day under the influence of stimulants
and warmth, he sank in the evening.

The patient in Case VII. was almost moribund on
admission to hospital. There was complete atony of
the bladder, and the condition of the mucous mem
brane of the bladder may be imagined from the fetid
state of the urine. The man was so weak and
exhausted that I could not entertain the idea of
performing lithotomy, so I gave him the only prospect
of recovery in performing litholapaxy. In Case XXIV.
death resulted from peritonitis, evidently produced by
extension of inflammation from the neck of the bladder.
It is difficult to say if this was due to the operation or
to impaction of the calculus at the neck of the bladder
and the force necessary to displace it. I am inclined
to think the latter was the cause. There was no
difficulty whatever in performing the operation; the

stone was very small, and the operation lasted only eight minutes. The bladder was healthy except at the neck, where the calculus had lain impacted for two days. The kidneys also were healthy. It is probable that the calculus, forced on by the accumulation of urine and efforts at micturition, acted as an irritant to the mucous membrane, and that the force necessary to push back the stone into the bladder increased the irritation, thereby producing inflammation. In Case XLIV. the cause of death was pyæmia, the result of cystitis supervening on the operation, and caused by the irritation of a large fragment, which it was found impossible to dispose of at the first sitting owing to the exhausted state of the patient. Death in Case XCVII. resulted apparently from the shock of the operation and exhaustion produced by the chloroform acting on a constitution worn out by the painful nature of the malady and extensive disease of the bladder, ureters, and kidneys. Emboldened by having previously removed a larger calculus successfully, in Case XVI.—the details of which have already been given—and by a long-continued series of successful cases, from Case XLV. to Case XCVI. inclusive, in many of them the calculi being very large, I had no hesitation in attacking the stone in this case ; and I feel confident the patient would have recovered but for the extensive disease of the kidneys which existed, and from which he must have succumbed to any operation. A large and powerful lithotrite sent me by Weiss worked admirably in this case. The manner in which a large oxalate of lime calculus, weighing over 3 ounces, and as hard as flint, was crushed to atoms and completely removed by the aspirator in fifty-two minutes, without the necessity of any dilatation of the urethra or injury to the urinary passage, shows what can be done by the new operation. Though the crushing power of this instrument is very great and the strength of the steel enormous, it was noticed, both by the surgeons who honoured me with their presence during the operation and by myself, that, when dealing with a large and hard stone of this kind, the leverage in the instrument is deficient. This might be easily remedied by having long knobs to the wheel-shaped handle of the male blade, and by the addition of a light cross-bar to the cylindrical handle of the female blade, both the knobs and cross-bar being removable, so that they might be applied only when large calculi are dealt with. In Case CXX, death, I regret to record, resulted from the neglect of the patient's relatives and the hospital subordinate, and was not justly attributable to the operation.

With the exception of the five deaths above described, all the cases made excellent recoveries. Cystitis, which so often followed on the old operation of litho-trity, and which often left the patients in almost as bad a condition as they were in before the operation, seldom occurs after litholapaxy ; and when it does occur, or had existed previous to the operation, is very amenable to treatment.

Should cystitis supervene on the operation, the treatment during the acute stage will consist in perfect rest in bed, the administration of alkalies with demul-cent drinks, hot baths, hot fomentations to the

hypogastric region, poultices, &c. The local treatment
that I have found most effectual in the subacute and
chronic stages of cystitis consists in the injection into
the bladder once daily of a weak astringent solution

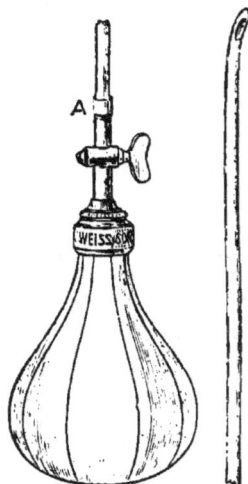

of either nitrate of silver or acetate
of lead. These astringent solutions
are best applied by means of an
eight - ounce india-rubber bottle
(Fig. 21), provided with a stop-
cock, the nozzle (A) of which fits
on to the end of a No. 10 or 12
soft gum-elastic catheter provided
with a large oval eye. The process
should be commenced by washing
out the bladder with warm water,
to which may be added a trace of
carbolic acid or a few drops of a
solution of permanganate of potash.

FIG. 21.

When the bladder has been thus cleansed of mucus.
pus, and shreds of lymph, the astringent is applied.
The astringent solution should at first be very mild,
say $\frac{1}{4}$ grain of either of the salts above mentioned to
the ounce, and may be gradually increased in strength,
no pain or uneasiness being produced.

These remarks do not, however, apply to those cases
of phosphatic cystitis sometimes met with in which
the chronic disease of the bladder is the *cause* of the
formation of soft phosphatic calculi. not the *result*.
In such cases the removal of the calculus will
not cure the disease. But the habitual use of the
india-rubber bottle and catheter for washing out the

bladder with weak astringents will alleviate the symptoms.

It sometimes happens, when the patient seems to be doing well for the first day or two after the operation, passing urine quite clear and troubled with no urinary symptoms, that on the third or fourth day the urine becomes quite milky, and slight irritability of the bladder occurs, the symptoms passing away in the course of a few days. Sir Henry Thompson was the first to call attention to this phenomenon,* and was unable then to account for its occurrence. I believe that I have read in some of his subsequent writings that he had come to the conclusion that it was due to infection carried from one bladder to another by means of the lithotrite. I am of opinion that the appearances referred to are due, as a rule, to nothing more or less than an exfoliation of the superficial layers of the mucous membrane of the bladder, the result of the comparatively rough treatment that it undergoes during the operation. It may be compared to the exfoliation of skin from the back of the hands with which Europeans in India are not unfamiliar who have exposed themselves to the scorching rays of the sun in the hot weather during a shooting trip into the jungle.

Litholapaxy was performed by me in four instances on females—three children and one adult—with the happiest results. The patient, in each instance, was going about the day after the operation, suffering no inconvenience whatever. The operation is compara-

---

* "Diseases of the Urinary Organs," sixth edition, p. 87.

tively easy of performance in females, owing to the urethra being so short and dilatable. No forcible dilatation of the parts was necessary, and, consequently, no incontinence of urine ensued—that troublesome sequel which so often results from the operation by dilatation.

Lithotomy in the male child, when skilfully performed, has always been recognized as a successful operation. It appears to me that litholapaxy is unsuited to such cases, owing to the undeveloped condition of the genito-urinary organs, the bladder being small, the urethra narrow, and the mucous membrane of both extremely sensitive and liable to laceration. It must be remembered that, in order that the aspirator may be of any practical utility for removing fragments, a much larger canula must be used than could with safety be introduced through the urethra of a male child. I am, however, unable to speak from practical experience in the matter, never having attempted litholapaxy in a male below the age of puberty ; nor have I any inclination to do so whilst my success from lithotomy in such cases continues to be what it has hitherto been.

Since the last paragraph was first published in the *Lancet*, Surgeon-Major Keegan, Bengal Medical Service, has published * two very able and interesting papers giving details of a considerable number of cases in which litholapaxy was performed by him in male children with remarkable success. He has undoubt-

* *Indian Medical Gazette*, June and September 1885.

edly shown that the operation may be successfully
performed in male children when the stone is small.
And, should litholapaxy in male children ever become
the rule, to Dr. Keegan must be assigned the credit of
being the first to show that the operation is feasible in
such cases.

I must confess that, since reading these papers by
Dr. Keegan, my views on the subject of litholapaxy
in children are considerably modified. That the operation is admissible in the case of small calculi in male
children there can be no question. The mortality
attending any operation must, however, be the first
consideration, and it has yet to be demonstrated that
the results obtained from litholapaxy in the case of
children are such as to justify that operation taking
the place of lithotomy—which has always been regarded
as a very successful operation in such cases—to any
large extent.

CONCLUDING REMARKS.

To sum up, then, it will be observed that, as the rule, lithotomy is the operation advocated in male children, and litholapaxy that in adult males and females of all ages. The two operations should be regarded as supplementary to each other, each having its own sphere. It is by regarding them as such that the best results will be obtained.

Though litholapaxy should be the rule in the treatment of stone in the adult, there must always be a small number of cases in which the operation will be inapplicable. Such cases are those in which the stone is extremely large and hard ; or where narrow, cartilaginous stricture of the urethra, great enlargement of the prostate with tortuous and pouched urethral canal, or contracted, rigid, and irritable bladder complicates the case. In these, lithotomy of some kind must still remain the proper operation. As the surgeon gains experience, however, and with further improvements in the instruments used, the number of cases in which litholapaxy will be unsuitable will gradually diminish. Several of the cases of stone in the adult treated by me by lithotomy after I first began to practise litholapaxy I would now treat by the latter operation.

The old operation of lithotrity may now be regarded
as dead and buried, and is not likely to be revived.
Under no circumstances can I conceive its practice
justifiable ; that is to say, with modern appliances at
hand, under no circumstances should a stone be crushed
and its fragments allowed to remain in the bladder to
come away by natural efforts.

In concluding this monograph, I would venture to
suggest that the time has now come when Bigelow's
operation should be universally recognized to be, what
it undoubtedly is, a distinctly new operation.   Shortly
after the introduction of litholapaxy, Sir Henry
Thompson assumed a position of opposition to this view.
In a lecture delivered at University College Hospital
in December 1878,* an attempt was made by him to
show that the new operation had been gradually
developed out of the old operation of lithotrity. the
previous existence of Clover's syringe, and the assertion
that Sir Henry himself had, during the previous two
years, been in the habit of doing more at each sitting.
both in the way of crushing and removal of fragments.
than formerly, being mainly relied on as the connecting
links in establishing their identity.   Bigelow's share in
the development of the operation was minimized : his
instruments were denounced as incapable of performing
the work assigned to them, and held up to ridicule as
" enormous and unwieldy "—suggesting to Sir Henry's
mind " some resuscitated relics of the early history of
lithotrity," reminding him of " the terrible engines

* *Lancet*, vol. i. 1879, p. 145.

used by Heurteloup "—and disastrous results were anticipated from the alleged proposal of Bigelow " to make the rule absolute to remove at one sitting an entire stone, no matter how large it may be or what the condition of the patient," a proposal which would seem to have had its origin in the imagination of the lecturer, for Bigelow asserts * that no such proposal had ever been made by him. Read by the light of six years' practical experience of the operation all over the world, the gloomy anticipations then expressed do not seem to have been realized. Sir Henry seems to have altered his opinions very materially since that time, for we find from his most recent writings that lithotrites and evacuating catheters, which were then pronounced dangerous and unnecessary, are now held to be admissible, and even necessary, when dealing with large calculi.

Now, every one who knows anything about the history of lithotrity must be aware that previous to the appearance of Bigelow on the scene, in 1878, the tendency of all lithotritists was (1) to restrict to the lowest possible limit the time occupied at each sitting, four or five minutes being the utmost time allowed as safe ; (2) to employ instruments of the smallest size possible ; and (3) to leave the evacuation of the fragments as much as possible to natural efforts. The principles on which this practice was founded were : (1) That the bladder was extremely intolerant of the presence of instruments ; and (2) that in direct propor-

* *Lancet*, vol. i. 1879, p. 693.

tion with the length of time instruments were manipulated there was the prospect of evil consequences resulting. There can be no doubt as to the teaching that prevailed on the subject. A reference to the latest editions of all the ordinary text-books published prior to 1879 will show that the authors were unanimous on these points ; and there was no one who inculcated the principles and practice referred to more strongly than Sir Henry Thompson himself, as might be illustrated by numerous extracts from his writings. It was reserved for Professor Bigelow to show that these principles were altogether wrong, and to introduce a practice entirely at variance with the old proceeding. The hypotheses on which the new operation was based were : (1) That the bladder was much more tolerant of prolonged manipulation than was previously supposed, and (2) that the temporary manipulation of blunt and polished instruments in the bladder was less irritating than the continued presence in the organ of sharp fragments of calculus. For the purpose of working out his idea, Bigelow introduced larger lithotrites than had previously been used, and invented an entirely new evacuating apparatus by which débris might be rapidly extracted from the bladder. In the introduction of the large evacuating canulæ, Bigelow availed himself of the discovery of Otis that the urethra is much more capacious than was previously recognized.

It is rather strange that Sir Henry Thompson should claim for Clover's syringe an efficiency as an aspirator which in its original and unmodified form it never possessed. The apparatus is referred to by most authors as a pretty and ingenious one for washing out the bladder. Its use in lithotrity is, however, deprecated, save in exceptional cases, such as where enlargement of the prostate or atony of the bladder co-exists ; and then the only efficiency claimed for it is that of washing out sand. Thus Mr. Cadge, of Norwich, writes : * "In doing this [removing calculi under the circumstances above referred to] I have sometimes used Clover's syringe, but more frequently have trusted to the quicker and less disturbing action of the scoop lithotrite." In 1869 Sir Henry Thompson, writing of the removal of fragments by Clover's syringe,† says : "The process is rather trying, however, for the bladder ; and it costs rather more pain and time than an ordinary sitting for lithotrity." Again, in 1871, he writes :‡ "Having used it [Clover's apparatus] very frequently, I would add that it is necessary to use all such apparatus with extreme gentleness, and I prefer to do without it, if possible." And that, even so late as 1878, Sir Henry relied much more on the flat-bladed lithotrite for the evacuation of the débris (Fergusson's method) than on Clover's syringe, is apparent from a passage in the lecture delivered by him in December 1878, already referred to.

Bigelow applied the name "litholapaxy" to his operation ; but to this Sir Henry Thompson objects, suggesting "lithotrity at one sitting" as more appropriate. Now, I think there are many advantages in

* *Lancet*, vol. i. 1879, p. 471.
† "Diseases of the Urinary Organs," p. 125.
‡ "Practical Lithotrity and Lithotomy," p. 215.

having a distinctly new name for a distinctly new
departure in surgery. The word litholapaxy (λίθος, a
stone, and λαπάξις, evacuation) seems to me the one
most expressive of the procedure involved in the new
operation. Bigelow's operation involves much more
than the crushing of the stone, the essential feature
being its complete and rapid evacuation. Besides, as
I have already pointed out, there are many cases in
which a small calculus can be removed by the aspirator
alone, in which no crushing is required, and to which,
consequently, the name " lithotrity at one sitting"
cannot be applied, whereas the word " litholapaxy "
will also embrace these.

There can be no doubt whatever that Bigelow's
operation was a distinct innovation both as regards the
principles involved and the means by which it was
accomplished. The operation struck at the root of all
previously held tenets regarding lithotrity ; and its introduction caused at the time considerable astonish-
ment to the profession all over the world. I must
confess my surprise that Sir Henry Thompson, after
employing Bigelow's operation in all its essential
details during the past five or six years, and obtaining
from it such brilliant results as those recorded by him,
should still persist in saying * that " no new form of
instrument is required by this operation," and that he
should refrain from according to Bigelow that credit to
which he is justly entitled for his originality.

Received at first with caution, the operation is

---

* Vide "Diseases of the Urinary Organs," seventh edition, p. 98, 1883.

steadily growing in favour with the profession in
America, England, and Europe generally. In India
litholapaxy has as yet been adopted by a few surgeons
only ; but I am convinced that the operation is destined
to play an important part in the surgery of the future
in this country, where unrivalled opportunities abound
for its practice, in reducing by a large percentage
the mortality, as well as the suffering, attendant
on stone in the bladder. I can testify to the
immense popularity of the modern operation amongst
the natives of India ; and it may be reasonably anticipated that, when it becomes generally known that a
small calculus may be removed from the bladder by an
operation that involves no cutting, little or no pain,
and confinement to hospital for a few days only, patients
will present themselves for treatment at an early stage
of the disease, when it is most amenable to treatment,
and when the operation is almost unattended with
danger.

The operation can no longer be said to be on its trial.
The prejudice against the operation that existed (and
still exists) in the minds of many surgeons, and which
is in great part due to the association of litholapaxy
with the old operation of lithotrity, from which it is
radically different, must gradually vanish before the
stern reality of facts such as those recorded in this
monograph. The surgeon who would give his patients
suffering from stone the best prospect of recovery
must practise litholapaxy. I believe I have pushed the
operation, as regards the size and hardness of the calculi
attacked, the ages and debilitated conditions of the

patients operated on, as far as any living surgeon, and
I cannot speak in terms too high of it.   By litholapaxy
the surgery of the bladder has been truly revolutionized ;
and I confidently anticipate that, with increased per-
fection in the instruments employed, larger calculi than
any hitherto attacked will successfully yield to the
operation.

INDEX.

PRINTED BY BALLANTYNE, HANSON AND CO.
LONDON AND EDINBURGH